Basics of
INTERFEROMETRY

Basics of
INTERFEROMETRY
Second Edition

P. HARIHARAN

School of Physics, University of Sydney
Sydney, Australia

AMSTERDAM • BOSTON • HEIDELBERG • LONDON
NEW YORK • OXFORD • PARIS • SAN DIEGO
SAN FRANCISCO • SINGAPORE • SYDNEY • TOKYO

ELSEVIER   Academic Press is an imprint of Elsevier

ACADEMIC
PRESS

Academic Press is an imprint of Elsevier
30 Corporate Drive, Suite 400, Burlington, MA 01803, USA
525 B Street, Suite 1900, San Diego, California 92101-4495, USA
84 Theobald's Road, London WC1X 8RR, UK

This book is printed on acid-free paper. ⊗

Library of Congress Cataloging-in-Publication Data

Hariharan, P.
Basics of interferometry / P. Hariharan.
    p. cm.
Includes bibliographical references and index.
ISBN-13: 978-0-12-373589-8 (hardcover : alk. paper)
ISBN-10: 0-12-373589-0 (hardcover : alk. paper) 1. Interferometry. I. Title.
QC411.H35 2006
535'.470287–dc22

                                        2006020075

British Library Cataloguing-in-Publication Data
A catalogue record for this book is available from the British Library.

ISBN 13: 978-0-12-373589-8
ISBN 10: 0-12-373589-0

For information on all Academic Press publications
visit our Web site at www.books.elsevier.com

Transferred to Digital Printing 2010

**Working together to grow
libraries in developing countries**

www.elsevier.com | www.bookaid.org | www.sabre.org

ELSEVIER   BOOK AID
            International   Sabre Foundation

To Raj

Contents

4 Source-Size and Spectral Effects 23

5 Multiple-Beam Interference 31

9

Optical Testing   67

10

Digital Techniques   83

13 Holographic and Speckle Interferometry 111

14 Interferometric Sensors 121

15 Interference Spectroscopy 133

16 Fourier Transform Spectroscopy   145

17 Interference with Single Photons   153

18 Building an Interferometer   165

A Monochromatic Light Waves   169

B Phase Shifts on Reflection   171

C Diffraction   173

D Polarized Light   177

E The Pancharatnam Phase   183

Preface to the First Edition

This book is intended as an introduction to the use of interferometric techniques for precision measurements in science and engineering. It is aimed at people who have some knowledge of optics but little or no previous experience in interferometry. Accordingly, the presentation has been specifically designed to make it easier for readers to find and assimilate the material they need.

The book can be divided into two parts. The first part covers such topics as interference in thin films and thick plates and the most common types of interferometers. This is followed by a review of interference phenomena with extended sources and white light, and multiple-beam interference. Laser light sources for interferometry and the various types of photodetectors are discussed.

The second part covers some important applications of optical interferometry: measurements of length, optical testing, studies of refractive index fields, interference microscopy, holographic and speckle interferometry, interferometric sensors, interference spectroscopy, and Fourier transform spectroscopy. The last chapter discusses the problems of setting up an interferometer, considers whether to buy or build one, and offers some suggestions.

Capsule summaries at the beginning and end of each chapter provide an overview of the topics explained in more detail in the text. Each chapter also contains suggestions for further reading and a set of worked problems utilizing real-world parameters that have been chosen to elucidate important or conceptually difficult questions.

Useful additional material is supplied in 15 appendices that cover the relevant aspects of wave theory, diffraction, polarization, and coherence, as well as related topics such as the Twyman–Green interferometer, the adjustment of the Mach–Zehnder interferometer, laser frequency shifting, heterodyne and phase-stepping techniques, the interpretation of shearing interferograms, holographic imaging, laser speckle, and laser frequency modulation by a vibrating surface.

This book would never have been completed without the whole-hearted support of several colleagues: in particular, Dianne Douglass, who typed most of the manuscript; Shirley Williams, who produced the camera-ready copy; Stuart Morris, who did many of the line drawings; Dick Rattle, who produced the photographs; and, last but not least, Philip Ciddor, Jim Gardner, and Kin Chiang, who reviewed the manuscript and made valuable suggestions at several stages. It is a pleasure to thank them for their help.

P. Hariharan
Sydney
April 1991

Preface to the Second Edition

The 15 years since the publication of the first edition of this book have seen an explosive growth of activity in the field of optical interferometry, including many new techniques and applications. The aim of this updated and expanded second edition is to provide an introduction to this rapidly growing field.

As before, the first part of the book covers basic topics such as interference in thin films and thick plates, the most common types of interferometers, interference phenomena with extended sources and white light, and multiple-beam interference. These topics are followed by a discussion of lasers as light sources for interferometry and the various types of photodetectors.

The second part covers some techniques and applications of optical interferometry. As before, the first four chapters deal with measurements of length, optical testing, studies of refractive index fields, and interference microscopy. A chapter has been added describing new developments in white-light interference microscopy. This chapter is followed by four chapters discussing holographic and speckle interferometry, interferometric sensors, interference spectroscopy, and Fourier transform spectroscopy. New sections have been added discussing recent developments such as gravitational wave detectors, optical signal processing, and laser frequency measurements. In view of the increasing importance of quantum optics, a new chapter on interference at the single-photon level has also been added. As before, the last chapter offers some practical suggestions on setting up an interferometer.

Some useful mathematical results as well as some selected topics in optics are summarized in 16 appendices, including new sections on Jones matrices and the Poincaré sphere and their use in visualizing the effects of retarders on polarized light, as well as the geometric (Pancharatnam) phase and its application to achromatic phase shifting.

I have used American spelling throughout in this book, except for the word "metre." Chapter 8 starts with a review of the work that led to the present standard of length based on the speed of light and cites the text of the internationally accepted definition of this unit. Accordingly, to avoid inconsistency, I have used the internationally accepted spelling for this word.

I am grateful to many of my colleagues for their assistance. In particular, I must mention Philip Ciddor, Maitreyee Roy, and Barry Sanders; without their help, this book could not have been completed.

P. Hariharan
Sydney
June 2006

Acknowledgments

I would like to thank the publishers, as well as the authors, for permission to reproduce the figures listed below:

American Physical Society (Figures 14.7, 15.5, 17.3, 17.4, 17.7), *Europhysics Letters* (Figure 17.2), Hewlett-Packard Company (Figure 8.3), *Japanese Journal of Applied Physics* (Figure 9.8), *Journal of Modern Optics* (Figures 17.5, 17.6, E.1), *Journal de Physique et le Radium* (Figure 16.1), Newport Corporation (Figure 18.1), North-Holland Publishing Company (Figures 9.11, 9.14, 13.5, 13.8), Penn Well Publishing Company (Figure 11.6), SPIE (Figures 9.7, 9.15), The Institute of Electrical and Electronics Engineers (Figure 14.3), The Institute of Physics (Figures 9.12, 14.4), The Optical Society of America (Figures 8.4, 10.3, 10.4, 11.2, 11.3, 11.8, 12.2, 12.3, 12.4, 13.1, 14.5, 15.3).

1

Introduction

Phenomena caused by the interference of light waves can be seen all around us: typical examples are the colors of an oil slick or a thin soap film.

Only a few colored fringes can be seen with white light. As the thickness of the film increases, the optical path difference between the interfering waves increases, and the changes of color become less noticeable and finally disappear. However, with monochromatic light, interference fringes can be seen even with quite large optical path differences.

Since the wavelength of visible light is quite small (approximately half a micrometre for green light), very small changes in the optical path difference produce measurable changes in the intensity of an interference pattern. As a result, optical interferometry permits extremely accurate measurements.

Optical interferometry has been used as a laboratory technique for almost a hundred years. However, several new developments have extended its scope and accuracy and have made the use of optical interferometry practical for a very wide range of measurements.

The most important of these new developments was the invention of the laser. Lasers have removed many of the limitations imposed by conventional light sources and have made possible many new interferometric techniques. New applications have also been opened up by the use of single-mode optical fibers to build analogs of conventional interferometers. Yet another development that has revolutionized interferometry has been the increasing use of photodetectors and digital electronics for signal processing. Interferometric measurements have also assumed increased importance with the redefinition of the international standard of length (the metre) in terms of the speed of light.

Some of the current applications of optical interferometry are accurate measurements of distances, displacements, and vibrations; tests of optical systems; studies of gas flows and plasmas; studies of surface topography; measurements of temperature, pressure, and electrical and magnetic fields; rotation sensing; high-resolution spectroscopy, and laser frequency measurements. Applications being explored include high-speed all-optical logic and the detection of gravitational waves. There is little doubt that, in the near future, many more will be found.

2

Interference: A Primer

In this chapter, we will discuss some basic concepts:

- Light waves
- Intensity in an interference pattern
- Visibility of interference fringes
- Interference with a point source
- Localization of interference fringes

2.1 LIGHT WAVES

Light can be thought of as a transverse electromagnetic wave propagating through space. Because the electric and magnetic fields are linked to each other and propagate together, it is usually sufficient to consider only the electric field at any point; this field can be treated as a time-varying vector perpendicular to the direction of propagation of the wave. If the field vector always lies in the same plane, the light wave is said to be linearly polarized in that plane. We can then describe the electric field at any point due to a light wave propagating along the z direction by the scalar equation

$$E(x, y, z, t) = a \cos[2\pi(\nu t - z/\lambda)], \qquad (2.1)$$

where a is the amplitude of the light wave, ν is its frequency, and λ is its wavelength. Visible light comprises wavelengths from 0.4 μm (violet) to 0.75 μm (red), corresponding, roughly, to frequencies of 7.5×10^{14} Hz and 4.0×10^{14} Hz, respectively. Shorter wavelengths lie in the ultraviolet (UV) region, while longer wavelengths lie in the infrared (IR) region.

The term within the square brackets, called the phase of the wave, varies with time as well as with the distance along the z-axis from the origin. With the passage of time, a surface of constant phase (a wavefront) specified by Eq. 2.1 moves along the z-axis with a speed

$$c = \lambda \nu \tag{2.2}$$

(approximately 3×10^8 metres per second in a vacuum). In a medium with a refractive index n, the speed of a light wave is

$$v = c/n \tag{2.3}$$

and, since its frequency remains unchanged, its wavelength is

$$\lambda_n = \lambda/n. \tag{2.4}$$

If a light wave traverses a distance d in such a medium, the equivalent optical path is

$$p = nd. \tag{2.5}$$

Equation 2.1 can also be written in the compact form

$$E(x, y, z, t) = a \cos[\omega t - kz], \tag{2.6}$$

where $\omega = 2\pi \nu$ is the circular frequency, and $k = 2\pi/\lambda$ is the propagation constant.

Equation 2.6 is a description of a plane wave propagating through space. However, with a point source of light radiating uniformly in all directions, a wavefront would be an expanding spherical shell. This leads us to the concept of a spherical wave, which can be described by the equation

$$E(r, t) = (a/r) \cos[\omega t - kr]. \tag{2.7}$$

At a very large distance from the source, such a spherical wave can be approximated, over a limited area, by a plane wave.

While the representation of a light wave, in terms of a cosine function, that we have used so far is easy to visualize, it is not well adapted to mathematical manipulation. It is often more convenient to use a complex exponential representation and write Eq. 2.6 in the form (see Appendix A)

$$E(x, y, z, t) = \mathrm{Re}\{a \exp(-i\phi) \exp(i\omega t)\}$$
$$= \mathrm{Re}\{A \exp(i\omega t)\}, \tag{2.8}$$

where $\phi = 2\pi z/\lambda$, and $A = a \exp(-i\phi)$ is known as the complex amplitude.

2.2 INTENSITY IN AN INTERFERENCE PATTERN

When two light waves are superposed, the resultant intensity at any point depends on whether they reinforce or cancel each other. This is the well-known phenomenon of interference. We will assume that the two waves are propagating in the same direction and are polarized with their field vectors in the same plane. We will also assume that they have the same frequency.

The complex amplitude at any point in the interference pattern is then the sum of the complex amplitudes of the two waves, so that we can write

$$A = A_1 + A_2, \tag{2.9}$$

where $A_1 = a_1 \exp(-i\phi_1)$ and $A_2 = a_2 \exp(-i\phi_2)$ are the complex amplitudes of the two waves. The resultant intensity is, therefore,

$$
\begin{aligned}
I &= |A|^2 \\
&= (A_1 + A_2)(A_1^* + A_2^*) \\
&= |A_1|^2 + |A_2|^2 + A_1 A_2^* + A_1^* A_2 \\
&= I_1 + I_2 + 2(I_1 I_2)^{1/2} \cos \Delta\phi,
\end{aligned} \tag{2.10}
$$

where I_1 and I_2 are the intensities due to the two waves acting separately, and $\Delta\phi = \phi_1 - \phi_2$ is the phase difference between them.

If the two waves are derived from a common source, so that they have the same phase at the origin, the phase difference $\Delta\phi$ corresponds to an optical path difference

$$\Delta p = (\lambda/2\pi)\Delta\phi, \tag{2.11}$$

or a time delay

$$\tau = \Delta p/c = (\lambda/2\pi c)\Delta\phi. \tag{2.12}$$

The interference order is

$$N = \Delta\phi/2\pi = \Delta p/\lambda = \nu\tau. \tag{2.13}$$

If $\Delta\phi$, the phase difference between the beams, varies linearly across the field of view, the intensity varies cosinusoidally, giving rise to alternating light and dark bands, known as interference fringes. These fringes correspond to loci of constant phase difference (or, in other words, constant optical path difference).

2.3 VISIBILITY OF INTERFERENCE FRINGES

The intensity in an interference pattern has its maximum value

$$I_{max} = I_1 + I_2 + 2(I_1 I_2)^{1/2}, \tag{2.14}$$

when $\Delta\phi = 2m\pi$, or $\Delta p = m\lambda$, where m is an integer, and its minimum value

$$I_{min} = I_1 + I_2 - 2(I_1 I_2)^{1/2}, \tag{2.15}$$

when $\Delta\phi = (2m + 1)\pi$, $\Delta p = (2m + 1)\lambda/2$.

The visibility $\mathcal{V}$ of the interference fringes is then defined by the relation

$$\mathcal{V} = \frac{I_{max} - I_{min}}{I_{max} + I_{min}}, \tag{2.16}$$

where $0 \leqslant \mathcal{V} \leqslant 1$. In the present case, from Eqs. 2.14 and 2.15,

$$\mathcal{V} = \frac{2(I_1 I_2)^{1/2}}{I_1 + I_2}. \tag{2.17}$$

2.4 INTERFERENCE WITH A POINT SOURCE

Consider, as shown in Figure 2.1, a transparent plate illuminated by a point source of monochromatic light, such as a laser beam brought to a focus. Interference takes place between the waves reflected from the front and back surfaces of the plate. These waves can be visualized as coming from the virtual sources S_1 and S_2, which are mirror images of the original source S. Interference fringes are seen on a screen placed anywhere in the region in which the reflected waves overlap.

With a plane-parallel plate (thickness d, refractive index n), a ray incident at an angle θ_1, as shown in Figure 2.2, gives rise to two parallel rays.

The optical path difference between these two rays is

$$\Delta p = 2nd \cos\theta_2 + \lambda/2, \tag{2.18}$$

where θ_2 is the angle of refraction within the plate (note the additional optical path difference of $\lambda/2$ introduced by reflection at one surface; see Appendix B). Since the optical path difference depends only on the angle of incidence θ_1, the interference fringes are, as shown in Figure 2.3(a), circles centered on the normal to the plate (fringes of equal inclination, or Haidinger fringes).

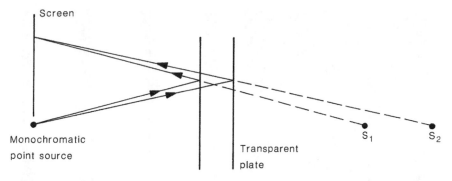

Figure 2.1. Interference with a monochromatic point source. Formation of interference fringes by the beams reflected from the two faces of a transparent plate.

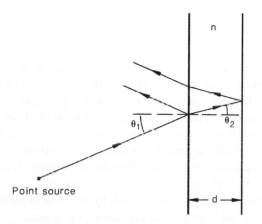

Figure 2.2. Interference with a monochromatic point source and a plane-parallel plate.

With a wedged plate and a collimated beam, the angles θ_1 and θ_2 are constant over the whole field, and the interference fringes are, as shown in Figure 2.3(b), contours of equal thickness (Fizeau fringes).

2.5 LOCALIZATION OF FRINGES

When an extended monochromatic source (such as a mercury vapor lamp with a monochromatic filter) is used, instead of a monochromatic point source, interference fringes are usually observed with good contrast only in a particular region. This phenomenon is known as localization of the fringes and is related to the lack of spatial coherence of the illumination.

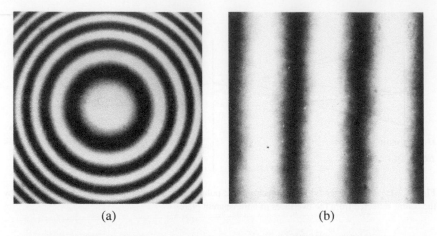

(a)   (b)

Figure 2.3.   (a) Fringes of equal inclination, and (b) fringes of equal thickness.

We will study the effects of coherence in more detail in Chapter 4. For the present, it is enough to say that such an extended source can be considered as an array of independent point sources, each of which produces a separate interference pattern. If the optical path differences at the point of observation are not the same for waves originating from different points on the source, these elementary interference patterns will, in general, not coincide and, when they are superposed, will produce an interference pattern with reduced visibility. It can be shown that the region where the visibility of the fringes is a maximum (the region of localization of the interference fringes) corresponds to the locus of points of intersection of pairs of rays derived from a single ray leaving the source.

Two cases are of particular interest. With a plane-parallel plate, as we have seen earlier, any incident ray gives rise to two parallel rays that meet only at infinity. Accordingly, the interference fringes (fringes of equal inclination) formed with an extended quasi-monochromatic source are localized at infinity. If the fringes are viewed through a lens, as shown in Figure 2.4, they are localized in its focal plane.

In the case of a wedged thin film, as shown in Figure 2.5, a ray from a point source S gives rise to two reflected rays that meet at a point P. Accordingly, with an extended source at S, the visibility of the interference pattern will be a maximum near P. In this case, the position of the region of localization of the interference fringe depends on the direction of illumination and can shift from one side of the film to the other. However, for near-normal incidence, the interference fringes are localized in the film. To a first approximation, the interference fringes are then contours of equal thickness.

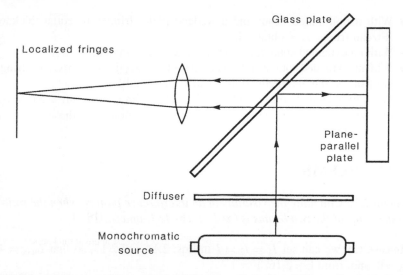

Figure 2.4. Interference with an extended light source. Formation of fringes of equal inclination localized at infinity.

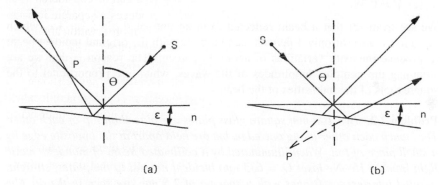

Figure 2.5. Formation of localized interference fringes by a thin wedged film.

2.6 SUMMARY

- If two beams derived from the same light source are superposed, a linear variation in the optical path difference produces a sinusoidal variation in the intensity (interference fringes).
- With a point source, interference fringes can be seen anywhere in the region where the beams overlap.
- With a point source and a plane-parallel plate, these interference fringes are fringes of equal inclination (Haidinger fringes).

- With a collimated beam and a wedged plate, fringes of equal thickness (Fizeau fringes) are obtained.
- With an extended source, localized fringes are obtained.
- With an extended source and a plane-parallel plate, these interference fringes are fringes of equal inclination, localized at infinity.
- With an extended source and a wedged thin film (see Figure 2.5), fringes of equal thickness are obtained. At near-normal incidence, these fringes are localized in the film.

2.7 PROBLEMS

Problem 2.1. *Calculate the visibility in an interference pattern when the ratio of the intensities of the two beams is (a) 1 : 1, (b) 4 : 1, and (c) 25 : 1.*

In case (a), we can set $I_1 = I_2 = I$ in Eqs. 2.14 and 2.15, so that $I_{max} = 4I$, $I_{min} = 0$, and, from Eq. 2.16, $\mathcal{V} = 1$.

In the other two cases, we can use Eq. 2.17. We then have

   (b) $\mathcal{V} = 0.8$,
   (c) $\mathcal{V} = 0.38$.

We see from (c) that a beam reflected from an untreated glass surface, which has a reflectance of only 4 percent, can interfere with the original incident beam to produce intensity variations of about 38 percent. The reason is that we are summing the complex amplitudes of the waves, which are proportional to the square roots of the intensities of the beams.

Problem 2.2. *Two 100-mm square glass plates are placed on top of each other. They touch each other along one edge, but are held apart at the opposite edge by a small piece of foil. When illuminated by a collimated beam of monochromatic light from an He–Ne laser (λ = 633 nm) incident normal to the plates, straight, parallel interference fringes with a spacing of 2.5 mm are seen in the air film between the plates. What is (a) the angle ε between the surfaces of the plates and (b) the thickness d of the foil?*

The interference fringes seen are fringes of equal thickness. From Eq. 2.18, the increase in the thickness of the air film from one fringe to the next is

$$\Delta d = \lambda/2 = 316.5 \text{ nm.} \tag{2.19}$$

The angle between the surfaces is, therefore,

$$\epsilon = \frac{316.5 \times 10^{-9}}{2.5 \times 10^{-3}} \text{ radian}$$

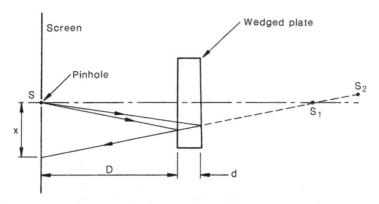

Figure 2.6. Formation of interference fringes with a monochromatic source and a wedged plate.

$$= 1.2666 \times 10^{-4} \text{ radian}$$

$$= 26.1 \text{ arc sec}, \qquad (2.20)$$

while the thickness of the foil is

$$d = 12.66 \ \mu\text{m}. \qquad (2.21)$$

Problem 2.3. *A plate of glass (thickness d = 3 mm, refractive index n = 1.53), whose faces have been worked flat and nominally parallel, is illuminated through a pinhole in a screen, as shown in Figure 2.6, by a point source of monochromatic light (λ = 633 nm). The plate is at a distance D = 1.00 m from the screen. The interference pattern seen on the screen is a set of concentric circles whose center lies at a distance x = 15 mm from the pinhole. What is the wedge angle between the faces of the plate?*

Interference takes place between the waves coming from the virtual sources S_1 and S_2, the images of S formed by reflection at the two faces of the glass plate. Nonlocalized circular fringes are formed with their center at the point at which the line joining S_1 and S_2 intersects the screen. To a first approximation, the angle between the two faces of the plate is[1]

$$\epsilon = \frac{xd}{2n^2 D^2}. \qquad (2.22)$$

---

[1] See J. H. Wasilik, T. V. Blomquist, and C. S. Willet, "Measurement of parallelism of the surfaces of a transparent sample using two-beam non-localized fringes produced by a laser," *Appl. Opt.* **10**, 2107–2112 (1971).

In the present case, we have

$$\epsilon = \frac{3 \times 15}{2 \times 1.53^2 \times 1000^2}$$

$$= 9.6 \times 10^{-6} \text{ radian}$$

$$= 2.0 \text{ arc sec.} \tag{2.23}$$

This simple test for parallelism can be carried out very quickly, since the position of the center of the pattern is not affected by small tilts of the glass plate. The sense of the wedge can be identified easily, because the center of the fringe pattern is always displaced toward the thicker end of the wedge.

FURTHER READING

For more information, see

1. E. Hecht, *Optics*, Addison-Wesley, Reading, MA (1987).
2. M. Born and E. Wolf, *Principles of Optics*, Cambridge University Press, Cambridge, UK (1999).
3. G. Brooker, *Modern Classical Optics*, Oxford University Press, Oxford, UK (2003).
4. P. Hariharan, *Optical Interferometry*, Academic Press, San Diego, CA (2003).

3

Two-Beam Interferometers

To make measurements using interference, we usually need an optical arrangement in which two beams traveling along separate paths are made to interfere. One of these paths is the reference path, while the other is the test, or measurement, path. The optical path difference between the interfering wavefronts is then

$$\Delta p = p_1 - p_2$$
$$= \Sigma(n_1 d_1) - \Sigma(n_2 d_2), \tag{3.1}$$

where n is the refractive index, and d the length, of each section in the two paths.

In order to produce a stationary interference pattern, the phase difference between the two interfering waves should not change with time. The two interfering beams must, therefore, have the same frequency. This requirement can be met only if they are derived from the same source.

Two methods are commonly used to obtain two beams from a single source. They are

- Wavefront division
- Amplitude division

3.1 WAVEFRONT DIVISION

Wavefront division uses apertures to isolate two beams from separate portions of the primary wavefront. In the configuration shown in Figure 3.1, used in Young's experiment to demonstrate the wave nature of light, the two pinholes can be regarded as secondary sources. Interference fringes are seen on a screen

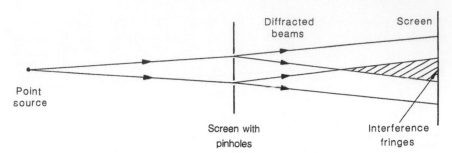

Figure 3.1.   Interference of two beams formed by wavefront division.

placed in the region of overlap of the diffracted beams from the two pinholes (see Appendix C).

Wavefront division is used in the Rayleigh interferometer (see Section 3.3).

3.2 AMPLITUDE DIVISION

In amplitude division, two beams are derived from the same portion of the original wavefront.

Some optical elements that can be used for amplitude division are shown in Figure 3.2.

The most widely used device is a transparent plate coated with a partially reflecting film that transmits one beam and reflects the other (commonly referred to as a beam splitter). A partially reflecting film can also be incorporated in a cube made up of two right-angle prisms with their hypotenuse faces cemented together.

Another device that has been used is a diffraction grating, which produces, in addition to the directly transmitted beam, one or more diffracted beams (see Appendix C).

Yet another device that can be used is a polarizing prism, which produces two orthogonally polarized beams. A polarizing beam splitter can also be constructed by incorporating in a beam-splitting cube a multilayer film which reflects one

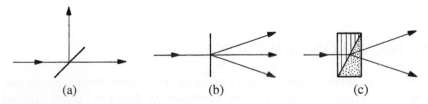

Figure 3.2.   Techniques for amplitude division: (a) a beam splitter, (b) a diffraction grating, and (c) a polarizing prism.

polarization and transmits the other. In both cases, the electric vectors must be brought back into the same plane, usually by means of another polarizer, for the two beams to interfere (see Appendix D).

Some common types of two-beam interferometers are

- The Rayleigh interferometer
- The Michelson (Twyman–Green) interferometer
- The Mach–Zehnder interferometer
- The Sagnac interferometer

3.3 THE RAYLEIGH INTERFEROMETER

The Rayleigh interferometer uses wavefront division to produce two beams from a single source. As shown in Figure 3.3, two sections of a collimated beam are isolated by a pair of apertures. The two beams are brought together in the focal plane of a second lens. Measurements are made on the interference pattern formed in this plane. Two identical glass plates are placed in the two beams, and the optical paths can be equalized by tilting one of them.

The Rayleigh interferometer has the advantages of simplicity and stability, and, since the two optical paths are equal at the center of the field, it is possible to use a white-light source. However, it has the disadvantage that the interference fringes are very closely spaced and must be viewed under high magnification; moreover, to obtain fringes with good visibility, a point or line source must be used (see Section 4.2).

The most common application of the Rayleigh interferometer is to measure the refractive index of a gas. When a gas is admitted into one of the evacuated tubes, the number of interference fringes crossing a fixed point in the field is given by the relation

$$N = \frac{(n-1)d}{\lambda},$$
(3.2)

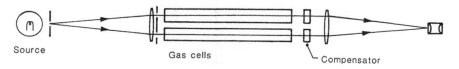

Source   Gas cells   Compensator

Figure 3.3. The Rayleigh interferometer.

where n is the refractive index of the gas, and d is the length of the tube. Measurements of the refractive index of a mixture of two gases can be used to determine its composition.

3.4 THE MICHELSON INTERFEROMETER

In the Michelson interferometer, the beam from the source is divided, as shown in Figure 3.4, at a semireflecting coating on the surface of a plane-parallel glass plate. The same beam splitter is used to recombine the beams reflected back from the two mirrors.

To obtain interference fringes with a white-light source, the two optical paths must be equal for all wavelengths. Both arms must, therefore, contain the same thickness of glass having the same dispersion. However, one beam traverses the beam splitter three times, while the other traverses it only once. Accordingly, a compensating plate (identical to the beam splitter, but without the semireflecting coating) is introduced in the second beam.

As shown in Figures 3.4 and 3.5, reflection at the beam splitter produces a virtual image M_2' of the mirror M_2. We can visualize the interfering beams as coming from the virtual sources S_1 and S_2, which are images of the original source S in M_1 and M_2'. The interference pattern observed is similar to that produced in a layer of air bounded by M_1 and M_2', and its characteristics depend on the nature of the light source and the separation of M_1 and M_2'.

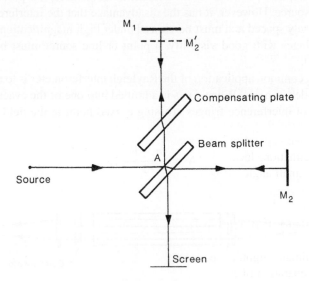

Figure 3.4.   The Michelson interferometer.

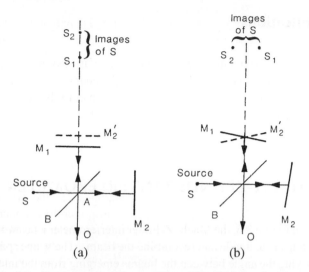

Figure 3.5. Formation of interference fringes in the Michelson interferometer.

3.4.1 Fringes Formed with a Point Source

When, as shown in Figure 3.5(a), M_1 and M_2' are parallel, but separated by a finite distance, the interference fringes obtained are circles centered on the normal to the mirrors (fringes of equal inclination).

When M_1 and M_2' make a small angle with each other, the interference fringes obtained are, in general, a set of hyperbolas. However, when M_1 and M_2' overlap, as shown in Figure 3.5(b), the fringes seen near the axis are equally spaced, parallel, straight lines (fringes of equal thickness).

3.4.2 Fringes Formed with an Extended Source

With an extended source, the interference fringes are localized (see Section 2.5). When M_1 and M_2' are parallel, but separated by a finite distance, fringes of equal inclination, localized at infinity, are obtained, and when M_1 and M_2' overlap at a small angle, fringes of equal thickness, localized on the mirrors, are obtained.

3.4.3 Fringes Formed with Collimated Light

With collimated light, fringes of equal thickness are always obtained, irrespective of the separation of M_1 and M_2'. The Michelson interferometer modified to use collimated light is known as the Twyman–Green interferometer.

3.4.4 Applications

The Michelson (Twyman–Green) interferometer is easy to set up and align (see Appendix F). The two optical paths are well separated, and the optical path difference between the beams can be varied conveniently by translating one of the mirrors. Its applications include measurements of length (see Section 8.2) and optical testing (see Section 9.2).

3.5 THE MACH–ZEHNDER INTERFEROMETER

As shown in Figure 3.6, the Mach–Zehnder interferometer uses two beam splitters and two mirrors to divide and recombine the beams. The fringe spacing is controlled by varying the angle between the beams emerging from the interferometer. In addition, for any given angle between the beams, the position of the point of intersection of a pair of rays originating from the same point on the source can be controlled by varying the lateral separation of the beams. With an extended source, this makes it possible to obtain interference fringes localized in any desired plane.

The Mach–Zehnder interferometer has two attractive features. One is that the two paths are widely separated and are traversed only once; the other is that the region of localization of the fringes can be made to coincide with the test object, so that an extended source of high intensity can be used. However, adjustment of the interferometer is not easy (see Appendix G).

The Mach–Zehnder interferometer is widely used for studies of fluid flow, heat transfer, and the temperature distribution in plasmas (see Section 11.2).

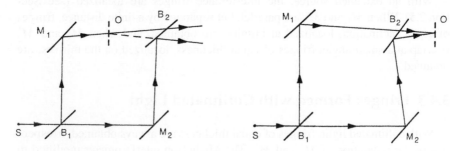

Figure 3.6. Localization of fringes in the Mach–Zehnder interferometer.

3.6 THE SAGNAC INTERFEROMETER

The Sagnac (pronounced *Sanyak*) interferometer is a common-path interferometer in which, as shown in Figure 3.7, the two beams traverse the same path in
opposite directions.

With most interferometers it is necessary to isolate the instrument from vibrations and air currents to obtain stable interference fringes. These problems are
much less serious with a common-path interferometer. In addition, since the optical paths traversed by the two beams in the Sagnac interferometer are very nearly
equal, interference fringes can be obtained immediately with an extended white-
light source.

Two forms of the Sagnac interferometer are possible, one with an even number
of reflections in each path [see Figure 3.7(a)], and the other with an odd number
of reflections in each path [see Figure 3.7(b)]. In the latter case, the wavefronts
are laterally inverted with respect to each other in some sections of the paths, so
that this form is not, strictly speaking, a common-path interferometer.

The Sagnac interferometer is extremely easy to align and very stable. Modified
versions of the Sagnac interferometer are widely used for rotation sensing instead
of conventional gyroscopes (see Section 14.4). Rotation of the interferometer with
an angular velocity Ω about an axis making an angle θ with the normal to the
plane of the beams introduces an optical path difference

$$\Delta p = (4\Omega A/c)\cos\theta \qquad (3.3)$$

between the two beams, where A is the area enclosed by the light paths.

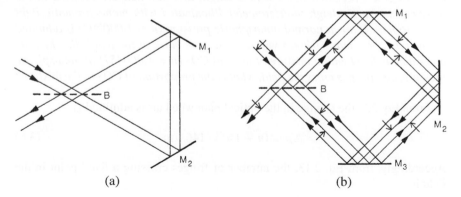

$$(a) \qquad\qquad\qquad\qquad (b)$$

Figure 3.7. Two forms of the Sagnac interferometer.

3.7 SUMMARY

Some common types of interferometers (and their applications) are

- The Rayleigh interferometer (gas analysis)
- The Michelson/Twyman–Green interferometer (length measurements/optical testing)
- The Mach–Zehnder interferometer (fluid flow)
- The Sagnac interferometer (rotation sensing)

3.8 PROBLEMS

Problem 3.1. *A Rayleigh interferometer uses collimating and imaging lenses with a focal length of 500 mm. The centers of the two apertures that define the beams are separated by 5 mm. If a white-light source is used with a narrow-band filter (mean wavelength 550 nm), what is the spacing of the interference fringes?*

At a distance x from the axis in the focal plane of the imaging lens, the additional optical path difference between the two interfering beams is

$$\Delta p = xa/f, \tag{3.4}$$

where a is the separation of the apertures, and f is the focal length of the lenses. Since successive maxima or minima in the interference pattern correspond to a change in the optical path difference of one wavelength, the separation of the interference fringes is

$$\Delta x = \lambda f/a = 0.055 \text{ mm}. \tag{3.5}$$

Problem 3.2. *Evacuated tubes with a length $d = 500$ mm are inserted in the two beams of a Rayleigh interferometer illuminated with monochromatic light ($\lambda = 546$ nm). If air at normal atmospheric pressure ($n = 1.000292$) is admitted to one tube, how many fringes will cross a fixed mark in the field? If, when the other tube is filled with a mixture of air and CO_2 ($n = 1.000451$) at atmospheric pressure, 282 fringes cross the field, what is the proportion of CO_2 in the mixture?*

From Eq. 2.5, the change in the optical path when air is admitted is

$$\Delta p = (n - 1)d = 146 \ \mu\text{m}. \tag{3.6}$$

Accordingly, from Eq. 2.13, the number of fringes crossing a fixed point in the field is

$$N = \Delta p/\lambda = 267.4. \tag{3.7}$$

The change in the optical path when the gas mixture is admitted is

$$\Delta p = 282\lambda = 154.0 \ \mu\text{m};$$ (3.8)

its refractive index is, therefore,

$$n = 1.000308.$$ (3.9)

Since the refractive index of the mixture is a linear function of the relative proportions of the two components, the mixture contains 10 percent CO_2.

Problem 3.3. *A Michelson interferometer has one of its mirrors mounted on a micrometer slide. When the interferometer is illuminated with monochromatic light ($\lambda = 632.8$ nm), and the screw of the micrometer is turned through one revolution, 1581 fringes cross a reference mark in the field. What is the pitch of the screw?*

The passage of each fringe corresponds to a displacement of the mirror of half a wavelength (316.4 nm). Accordingly, the pitch of the screw is

$$\Delta z = 1581 \times 316.4 \times 10^{-9} \ \text{m}$$

$$= 0.5002 \ \text{mm}.$$ (3.10)

Problem 3.4. *A Sagnac interferometer, in the form of a square with sides 3.0 m long, is set up on a carousel and illuminated with white light. How fast would the carousel have to rotate for a detectable shift of the fringes? How would you make sure that this is not a spurious effect?*

The minimum fringe shift that can be detected by the eye is about 0.1 of the fringe spacing, corresponding to the introduction of an optical path difference of 0.1 λ between the beams. With white light (mean wavelength 550 nm), this would require, from Eq. 3.3, an angular velocity

$$\Omega = 0.1 \times 550 \times 10^{-9} \times 3 \times 10^8 / 4 \times 9$$

$$= 0.458 \ \text{radian/sec},$$ (3.11)

which would correspond to a speed of rotation of 8.75 rpm. To verify that this fringe shift is not spurious, the direction of rotation of the carousel should be reversed; a shift of the fringes in the opposite direction, of equal magnitude, should be observed.

FURTHER READING

For more information, see

1. A. A. Michelson, *Light Waves and Their Uses*, University of Chicago Press, Chicago (1907).
2. C. Candler, *Modern Interferometers*, Hilger and Watts, London (1951).
3. W. H. Steel, *Interferometry*, Cambridge University Press, Cambridge (1983).

4

Source-Size and Spectral Effects

The simple theory of interference outlined in Chapter 2 is not adequate to cover various effects that are observed with some of the commonly used light sources. One such effect that we have already encountered is fringe localization. Some topics we will discuss in this chapter are

- Coherence
- Source-size effects
- Spectral effects
- Polarization effects
- White-light fringes
- Channeled spectra

4.1 COHERENCE

With a perfectly monochromatic point source, the variations of the electrical field at any two points in space are completely correlated. The light is then said to be *coherent*. However, light from a thermal source such as a mercury vapor lamp, even when it consists only of a single spectral line, is not strictly monochromatic. Both the amplitude and the phase of the electric field at any point on the source exhibit rapid, random fluctuations. For waves originating from different points on the source, these fluctuations are completely uncorrelated. As a result, the light from such a source is only partially coherent. The visibility of the interference fringes is then determined by the coherence of the illumination.

With monochromatic light, the correlation (see Appendix H) between the fields at any two points on a wavefront is a measure of the spatial coherence of the light and normally depends on the size of the source. With an extended source, the region of localization of the interference fringes corresponds to the locus of points of intersection of rays derived from a single point on the source and, therefore, to the region where the correlation between the interfering fields is a maximum. The extent of the region of localization of the fringes is, therefore, related to the spatial coherence of the illumination. In the same manner, the correlation between the fields at the same point, at different times, is a measure of the temporal coherence of the light and is related to its spectral bandwidth. The maximum value of the optical path difference at which interference fringes can be observed is, therefore, a measure of the temporal coherence of the illumination. A more detailed treatment of coherence is presented in Appendix I; we discuss some useful results in the next two sections.

4.2 SOURCE-SIZE EFFECTS

We consider first a situation in which effects due to the spectral bandwidth of the light (or, in other words, due to the fact that it is not strictly monochromatic) can be neglected. Typically, this is the situation when the light is very nearly monochromatic (see Section 6.3), or when the optical path difference is very small ($\Delta p < 1$ mm with a low-pressure mercury vapor lamp and a filter that transmits only the green line). With interferometers using amplitude division (in which interference takes place between light waves coming from the same point on the original wavefront), interference fringes with good visibility can be obtained, even with such an extended source. However, to obtain interference fringes with good visibility with a system using wavefront division, such as the Rayleigh interferometer (see Section 3.3), it is necessary to limit the extent of the source by means of a small aperture (a pinhole, or a slit with its length parallel to the interference fringes). Coherence theory (see Appendix I.5) can be used to calculate the maximum permissible diameter of the pinhole, or width of the slit.

4.2.1 Slit Source

The intensity distribution across a rectangular slit of width b is

$$I(x) = \text{rect}(x/b), \tag{4.1}$$

where $\text{rect}(x) = 1$ when $|x| \leqslant 1/2$, and 0 when $|x| > 1/2$.

If the separation of the centers of the two beams in the interferometer is a, and the slit is set with its long dimension parallel to the interference fringes, the visibility of the fringes is, from Eq. H.5,

$$\mathcal{V} = \text{sinc}(ab/\lambda f), \tag{4.2}$$

where f is the focal length of the imaging lens in Figure 3.3, and sinc $x = (1/\pi x)\sin(\pi x)$. As the width of the slit is increased, the visibility of the fringes decreases; it drops to zero when

$$b = \lambda f/a. \tag{4.3}$$

4.2.2 Circular Pinhole

With a circular pinhole of diameter d, the visibility of the fringes is, from Eq. H.6,

$$\mathcal{V} = 2J_1(u)/u, \tag{4.4}$$

where $u = 2\pi ad/\lambda f$. The visibility of the fringes drops to zero when

$$d = 1.22\lambda f/a. \tag{4.5}$$

4.3 SPECTRAL EFFECTS

The other limiting case is when the source is a point (or we use amplitude division, so that interference takes place between corresponding elements of the original wavefront) but radiates over a range of wavelengths. The visibility of the interference fringes then falls off as the optical path difference between the beams is increased. The maximum value of the optical path difference at which fringes can be seen (which corresponds to the coherence length of the radiation, as defined in Appendix I.7), for radiation with a spectral bandwidth $\Delta \nu$ or $\Delta \lambda$, is given approximately by the relation

$$\Delta p = c/\Delta \nu = \lambda^2/\Delta \lambda. \tag{4.6}$$

4.4 POLARIZATION EFFECTS

Two beams polarized in orthogonal planes cannot interfere, since their field vectors are at right angles to each other. Similarly, two beams circularly polarized in opposite senses cannot produce an interference pattern. Accordingly, for maximum visibility of the interference fringes, the two beams leaving an interferometer must be in identical states of polarization. If they are linearly polarized in planes making an angle θ with each other, the visibility of the fringes would be

$$\mathcal{V}_\theta = \mathcal{V}_0 \cos\theta, \tag{4.7}$$

where $\mathcal{V}_0$ is the visibility of the fringes when $\theta = 0$.

If the light entering an interferometer is unpolarized or partially polarized, it can be regarded as made up of two orthogonally polarized components. We can then use the Jones calculus (see Appendix D.3) to evaluate the changes in the states of polarization of the two beams from the point where they are divided to the point where they are recombined. These changes must be identical if the interferometer is to be compensated for polarization.

A simple case in which an interferometer is compensated for polarization is when the normals to all the beam splitters and mirrors are in the same plane. Compensation for polarization is not possible if the interferometer contains elements such as cube corners. In this case, two suitably oriented polarizers must be used, one at the input to the interferometer and the other at the output. It is then possible to bring the emerging beams into the same state of polarization and equalize their amplitude, so as to obtain interference fringes with good visibility.

4.5 WHITE-LIGHT FRINGES

With white light a separate fringe system is produced for each wavelength, and the resultant intensity at any point in the plane of observation is obtained by summing these individual patterns.

If an interferometer is adjusted so that the optical path difference is zero at the center of the field of view, all the fringe systems formed with different wavelengths will exhibit a maximum at this point. A white central fringe is obtained with a dark fringe on either side. However, because the spacing of the fringes varies with the wavelength, the fringes formed by different wavelengths will no longer coincide as we move away from the center of the pattern. The result is a sequence of colors whose saturation decreases rapidly. (Note: In the case of fringes formed in a thin air film between two glass plates, a dark fringe is obtained where the two plates touch each other; this is due to the additional phase shift of π introduced on reflection at one of the surfaces: see Appendix B.)

Observation of the central bright (or dark) fringe formed with white light can be used to adjust an interferometer so that the two optical paths are equal (zero interference order). The following procedure can be used to find the central zero-order fringe:

1. Equalize the paths approximately, by measurement.
2. With a monochromatic source, adjust the interferometer so that a few interference fringes are seen in the field of view.
3. With a source emitting over a fairly broad spectral band (for example, a fluorescent lamp), adjust the optical paths so that the contrast of the fringes is at maximum.
4. Replace the fluorescent lamp with a white-light source and locate the zero-order bright (or dark) fringe.

4.6 CHANNELED SPECTRA

With a white-light source, interference fringes cannot be seen with the naked eye for optical path differences greater than about 10 μm. However, if the light leaving the interferometer is allowed to fall on the slit of a spectroscope, interference effects can be observed with much larger optical path differences.

Consider a thin film (thickness d, refractive index n) illuminated normally by a collimated beam of white light. Those wavelengths that satisfy the condition

$$(2nd/\lambda) = m,\qquad(4.8)$$

where m is an integer, correspond to interference minima and will be missing in the reflected light. If the reflected light is allowed to fall on a spectroscope, as shown in Figure 4.1, the spectrum will be crossed by dark bands (channeled spectra) as shown in Figure 4.2, corresponding to wavelengths satisfying Eq. 4.8.

For two wavelengths, λ_1 and λ_2, corresponding to adjacent dark bands,

$$(2nd/\lambda_1) = m \quad \text{and} \quad 2nd/\lambda_2 = m + 1,\qquad(4.9)$$

so that

$$d = \frac{\lambda_1\lambda_2}{2n|\lambda_2 - \lambda_1|}.\qquad(4.10)$$

Channeled spectra can be used to measure the thickness of thin transparent films (5–20 μm thick).

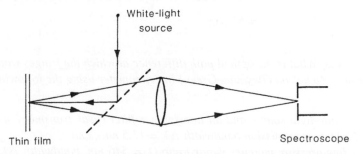

Figure 4.1. Arrangement for viewing the channeled spectrum formed by interference in a thin film.

Figure 4.2. Channeled spectrum produced by interference in a thin film.

Since the spacing of the bands increases as the optical path difference is reduced, observations of the channeled spectrum formed when an interferometer is illuminated with white light, using a pocket spectroscope, can help to speed up the adjustment of the interferometer for equal optical paths.

4.7 SUMMARY

To obtain interference fringes with good visibility with a thermal source:

- the source size must be small,
- the optical path difference must be small,
- both beams must have the same polarization.

White-light fringes (channeled spectra) can be used to equalize the optical paths in an interferometer and to measure the thickness of thin films.

4.8 PROBLEMS

Problem 4.1. *In the Rayleigh interferometer described in Problem 3.1, the source is a slit illuminated by a tungsten lamp with a narrow-band filter ($\lambda = 550$ nm). How far can the slit be opened before the interference fringes disappear?*

According to Eq. 4.3, the fringes will disappear when the width of the slit is

$$b = \lambda f / a$$
$$= 550 \times 10^{-9} \times 0.5/5 \times 10^{-3}$$
$$= 55 \ \mu\text{m}. \tag{4.11}$$

Problem 4.2. *What is the optical path difference at which the fringes would disappear in a Michelson (Twyman–Green) interferometer using the following light sources:*

(a) *a white-light source and an interference filter (peak transmission at $\lambda = 550$ nm, transmission bandwidth $\Delta\lambda = 11.5$ nm), and*

(b) *a low-pressure mercury vapor lamp ($\lambda = 546$ nm, bandwidth $\Delta\lambda = 5 \times 10^{-3}$ nm)?*

From Eq. 4.6, the values of the optical path difference at which the fringes disappear in the two cases are

$$\text{(a)} \ \Delta p = \left(550 \times 10^{-9}\right)^2 / 11.5 \times 10^{-9}$$
$$= 26.3 \ \mu\text{m}, \tag{4.12}$$

(b) $\Delta p = (546 \times 10^{-9})^2 / 5 \times 10^{-3} \times 10^{-9}$

$= 59.6$ mm. (4.13)

These values of the optical path difference correspond to the coherence length of the light in the two cases.

Problem 4.3. *If the same mercury vapor lamp is used as the source in a Fizeau interferometer (see Section 9.1), what would be the separation of the plates at which the interference fringes would disappear?*

In the Fizeau interferometer, the optical path difference between the beams is twice the separation of the plates. Accordingly, the fringes will disappear when the separation of the plates is 29.8 mm.

Problem 4.4. *The optical path difference between the beams in a Michelson interferometer is 10 mm. A high-pressure mercury vapor lamp is set up as the source and switched on. What happens to the interference fringes as the lamp warms up?*

As the lamp warms up, the pressure of the vapor filling increases. The velocity of the atoms increases, and the mean time between collisions decreases. Both these effects increase the width of the spectral line that is emitted. As a result, the visibility of the fringes decreases until, finally, they disappear.

Problem 4.5. *The two beams in an interferometer are linearly polarized and have equal intensities. If the angle between their planes of polarization is 60°, what is the visibility of the fringes?*

From Eq. 4.7, the visibility of the fringes is

$$\mathcal{V} = \cos 60°$$

$$= 0.5. \tag{4.14}$$

Problem 4.6. *With the setup shown in Figure 4.1, channeled spectra are observed in a plastic film (refractive index n = 1.47). Two adjacent dark bands are located at wavelengths of 0.500 μm and 0.492 μm. What is the thickness of the film?*

From Eq. 4.10, the thickness of the film is

$$d = 0.500 \times 0.492 / 2 \times 1.47 \times 0.008$$

$$= 10.46 \ \mu m. \tag{4.15}$$

FURTHER READING

For more information, see

1. M. Françon, *Optical Interferometry*, Academic Press, New York (1966).
2. P. Hariharan, *Optical Interferometry*, Academic Press, San Diego (2003).

5

Multiple-Beam Interference

When we studied the formation of interference fringes in plates and thin films in Section 2.4, we only considered the first reflection at each surface. With highly reflecting surfaces, we must take into account the effects of multiply reflected beams.

In this chapter we study

- Multiple-beam fringes by transmission
- Multiple-beam fringes by reflection
- Multiple-beam fringes of equal thickness
- Fringes of equal chromatic order (FECO fringes)
- The Fabry–Perot interferometer

5.1 MULTIPLE-BEAM FRINGES BY TRANSMISSION

Consider a light wave (unit amplitude) incident, as shown in Figure 5.1, on a plane-parallel plate (thickness d, refractive index n) at an angle θ_1. Multiple reflections at the surfaces of the plate produce a series of transmitted and reflected components, whose amplitudes fall off progressively.

The phase difference between successive transmitted, or reflected, components is

$$\phi = (4\pi/\lambda)nd\cos\theta_2, \tag{5.1}$$

where θ_2 is the angle of refraction within the plate.

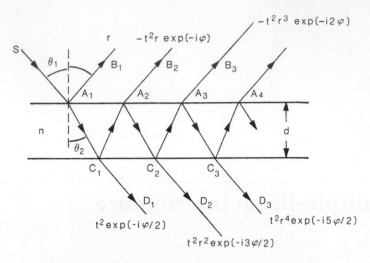

Figure 5.1. Multiple-beam interference in a plane-parallel plate.

The complex amplitude of the transmitted wave, which is the sum of the complex amplitudes of the transmitted components, is

$$A_T(\phi) = t^2\left[1 + r^2\exp(-i\phi) + r^4\exp(-i2\phi) + \cdots\right]$$
$$= t^2\big/\left[1 - r^2\exp(-i\phi)\right], \tag{5.2}$$

where r and t are, respectively, the coefficients of reflection and transmission (for amplitude) of the surfaces. The intensity in the interference pattern formed by transmission is, therefore,

$$I_T(\phi) = \left|A_T(\phi)\right|^2$$
$$= T^2\big/\left(1 + R^2 - 2R\cos\phi\right), \tag{5.3}$$

where $R = r^2$ and $T = t^2$ are, respectively, the reflectance and transmittance (for intensity) of the surfaces. The curves in Figure 5.2 show that as the reflectance R increases, the intensity at the minima decreases, and the bright fringes become narrower.

The separation of the fringes corresponds to a change in ϕ of 2π. The width of the fringes (Full Width at Half Maximum, or FWHM) is defined as the separation of two points, on either side of a maximum, at which the intensity is equal to half its maximum value. At these points,

$$\sin(\phi/2) = (1 - R)/2R^{1/2}. \tag{5.4}$$

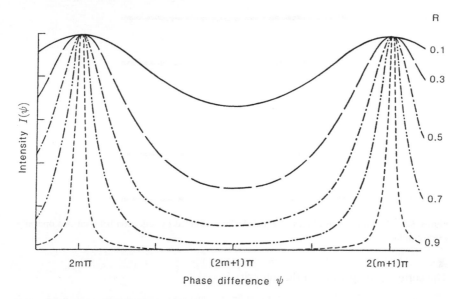

Figure 5.2. Intensity distribution in multiple-beam fringes formed by transmitted light, for different values of the reflectance (R) of the surfaces.

When R is close to unity, $\sin(\phi/2) \approx (\phi/2)$, and the change in ϕ corresponding to the FWHM of the fringes is

$$\Delta\phi_W = 4(1 - R)/2R^{1/2}. \tag{5.5}$$

The finesse of the fringes is defined as the ratio of the separation of adjacent fringes (corresponding to a change in ϕ of 2π) to their FWHM and is given by the relation

$$F = 2\pi/\Delta\phi_W = \pi R^{1/2}/(1 - R). \tag{5.6}$$

5.2 MULTIPLE-BEAM FRINGES BY REFLECTION

The complex amplitude of the reflected wave is obtained by summing the complex amplitudes of all the reflected components and is given by the relation

$$
\begin{aligned}
A_R(\phi) &= r\left[1 - t^2 \exp(-i\phi) - t^2 r^2 \exp(-i2\phi) + \cdots\right] \\
&= r\left[1 - \exp(-i\phi)\right]/\left[1 - r^2 \exp(-i\phi)\right].
\end{aligned} \tag{5.7}
$$

Figure 5.3. Multiple-beam fringes of equal thickness formed by reflection between two optically worked surfaces.

The corresponding value of the intensity is

$$I_R(\phi) = 2R(1 - \cos\phi)/(1 + R^2 - 2R\cos\phi). \qquad (5.8)$$

The interference fringes obtained by reflection are complementary to those formed by transmission; when $R \to 1$, very narrow dark fringes, on a bright background, are obtained.

5.3 MULTIPLE-BEAM FRINGES OF EQUAL THICKNESS

Multiple-beam fringes of equal thickness are much narrower than normal two-beam fringes and can be used to obtain a substantial improvement in accuracy when evaluating surface profiles. To obtain the best results, the angle between the two surfaces, and their separation, must be small. Figure 5.3 shows multiple-beam fringes of equal thickness formed by reflection in the wedged air film between two optically worked flat surfaces.

5.4 FRINGES OF EQUAL CHROMATIC ORDER (FECO FRINGES)

Multiple-beam fringes of equal chromatic order (FECO fringes) can be obtained with two highly reflecting surfaces enclosing a thin air film, using a white-light source and a setup similar to that employed for channeled spectra (see Section 4.6). Since the reflecting surfaces can be set parallel to each other, very nar-

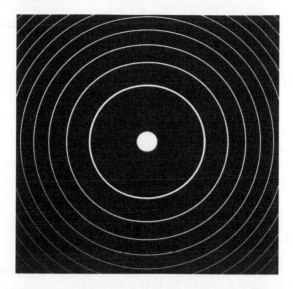

Figure 5.4. Fabry–Perot fringes obtained with a monochromatic source.

row, dark fringes can be obtained. FECO fringes have been used widely to study the microstructure of surfaces (see Section 11.4).

5.5 THE FABRY–PEROT INTERFEROMETER

The Fabry–Perot interferometer makes use of multiple-beam interference and consists, in its simplest form, of two parallel surfaces with semitransparent, highly reflecting coatings. If the separation of the surfaces is fixed, the instrument is commonly referred to as a Fabry–Perot etalon.

With an extended source of monochromatic light (wavelength λ), the interference pattern consists, as shown in Figure 5.4, of narrow, concentric rings (fringes of equal inclination) corresponding to the condition

$$2nd\cos\theta = m\lambda, \tag{5.9}$$

where d is the separation of the surfaces, n is the refractive index of the medium between them, θ is the angle of incidence within the interferometer, and m is an integer.

With a collimated beam at normal incidence, the transmittance of the interferometer exhibits sharp peaks at wavelengths defined by the condition

$$2nd = m\lambda. \tag{5.10}$$

The Fabry–Perot interferometer is widely used as a high-resolution spectrometer to study the fine structure of spectral lines (see Section 15.2).

5.6 SUMMARY

- Highly reflecting surfaces produce very sharp multiple-beam fringes.
- Narrow bright fringes, on a dark background, are obtained by transmission.
- Narrow dark fringes, on a bright background, are obtained by reflection.
- FECO fringes are used to study the microstructure of surfaces.
- The Fabry–Perot interferometer is used to study the fine structure of spectral lines.

5.7 PROBLEMS

Problem 5.1. *What is the finesse of multiple-beam fringes produced by two surfaces with a reflectance $R = $ (a) 0.8, (b) 0.9, (c) 0.95?*

From Eq. 5.6, the finesse of the fringes for these three values of the reflectance would be 14.1, 29.8, and 61.2, respectively.

Problem 5.2. *Local deviations from straightness of the order of a tenth of the width (FWHM) can be detected in multiple-beam fringes of equal thickness formed between two plates. What is the smallest surface step that can be detected with a monochromatic source whose wavelength is $\lambda = 546$ nm if the reference and test surfaces are coated so that their reflectance $R = 0.90$?*

From Eq. 5.6, the finesse of the fringes (the ratio of the spacing of the fringes to their FWHM) is 29.8. The increment in thickness from one fringe to the next is $\lambda/2 = 273$ nm. Accordingly, the smallest step that can be detected is

$$\Delta d = 0.1 \times 273 \times 10^{-9}/29.8$$

$$= 0.92 \text{ nm.} \tag{5.11}$$

FURTHER READING

For more information, see

1. S. Tolansky, *Multiple-Beam Interferometry of Surfaces and Films*, Clarendon Press, Oxford (1948).
2. S. Tolansky, *Surface Microtopography*, Longmans, London (1961).

3. G. Hernandez, *Fabry–Perot Interferometers*, Cambridge University Press, Cambridge, UK (1986).
4. J. M. Vaughan, *The Fabry–Perot Interferometer*, Adam Hilger, Bristol (1989).

6

The Laser as a Light Source

Several types of interferometers require a point source of monochromatic light. The closest approximation to such a source was, for many years, a pinhole illuminated by a mercury vapor lamp through a filter selecting a single spectral line. However, such a thermal source had two major drawbacks. One was the very small amount of light available; the other, as discussed in Chapter 4, was the limited spatial and temporal coherence of the light. The laser has eliminated these problems and provides an intense source of light with a remarkably high degree of spatial and temporal coherence. In this chapter we discuss

- Lasers for interferometry
- Laser modes
- Single-wavelength operation of lasers
- Polarization of laser beams
- Wavelength stabilization of lasers
- Laser-beam expansion
- Problems with laser sources
- Laser safety

6.1 LASERS FOR INTERFEROMETRY

Some lasers that are commonly used for interferometry are listed in Table 6.1.

Helium–neon (He–Ne) lasers are widely used for interferometry because they are inexpensive and provide a continuous, visible output. They normally operate at a wavelength of 633 nm, but modified versions are available with useful outputs at other visible and infrared wavelengths.

39

Table 6.1
Lasers for Interferometry

Laser type	Wavelength (μm)	Output
He–Ne	3.39, 1.15, 0.63, 0.61, 0.54	0.5–25 mW
Ar$^+$	0.51, 0.49, 0.35	0.5 W–a few W
Diode	1.064, 0.780, 0.660, 0.635, 0.594, 0.532, 0.475, 0.405	5–50 mW
Dye	1.08–0.41	10–100 mW
CO_2	~10.6, ~9.0	few W–few kW
Ruby	0.69	0.6–10 J
Nd-YAG	1.06	0.1–0.15 J

Argon-ion (Ar$^+$) lasers are more expensive but can provide much higher outputs. They can be operated at any one of a number of wavelengths in the visible and near UV regions, of which the strongest are those listed.

Laser diode systems now provide a range of wavelengths from the near infrared to the violet region of the spectrum. They are very compact and have a low power consumption. A drawback with diode lasers is that the output beam is divergent and astigmatic. However, packages are available incorporating additional optics to produce a collimated beam. Diode lasers can be tuned over a limited wavelength range by varying the injection current.

Dye lasers can be tuned over a range of 50–80 nm with any given dye. Operation at any wavelength in the visible region is possible by choosing a suitable dye.

Carbon dioxide lasers can be operated at a number of wavelengths in two bands in the infrared region. Because of their long output wavelength, they are useful for measurements over long distances.

Very short pulses of light (<20 ns duration), with very high peak powers, are produced by ruby and Nd-YAG lasers.

Shorter wavelengths can be produced with diode and solid-state lasers by using a nonlinear crystal as a frequency doubler.

6.2 LASER MODES

A laser basically consists of a section of an active medium, in which energy is stored, confined within a resonant cavity (a Fabry–Perot interferometer; see Section 5.5). In the He–Ne laser shown in Figure 6.1, the atoms in the gas mixture constituting this active medium are excited to higher energy levels by a DC discharge. A light wave with the right wavelength, originating anywhere within the cavity, will return in the correct phase along the same path, so that it draws energy

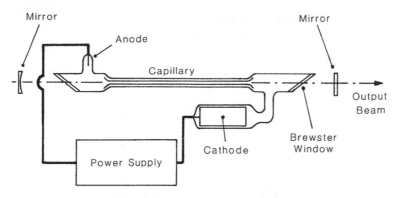

Figure 6.1. Schematic of a He–Ne laser.

from the active medium and is amplified at each pass through it. An output beam is obtained by making one mirror of the cavity a partially transmitting mirror.

The simplest form of resonant cavity uses two flat mirrors, but such a cavity is difficult to align and only marginally stable. A more stable configuration is one using two spherical mirrors whose separation is equal to their radius of curvature r, so that their foci coincide (a confocal cavity) (see Section 15.2.2).

With proper adjustment, only waves propagating parallel to the axis (the TEM_{00} mode) are amplified. However, the laser cavity can resonate at any wavelength satisfying the condition that the optical path for a round trip within the cavity is an integral number of wavelengths. For a confocal resonator, the frequencies of these longitudinal modes are given by the relation

$$\nu_m = (c/2L)(m + 1/2), \tag{6.1}$$

where L is the separation of the mirrors and m is an integer. The frequency difference between adjacent modes is, therefore,

$$\Delta\nu = c/2L. \tag{6.2}$$

If, as shown in Figure 6.2(a), more than one of these resonant wavelengths lies within the section of the gain profile in which the gain in the active medium exceeds the cavity losses, the laser may oscillate in several longitudinal modes corresponding to these wavelengths. The presence of more than one wavelength in the output of an He–Ne laser limits the coherence length to a few centimetres (see Appendix I.7).

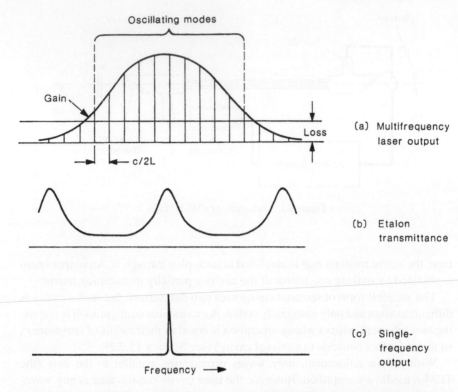

Figure 6.2.   (a) Frequency spectrum of a gas laser operating in multiple longitudinal modes; (b) transmission of an intracavity etalon, and (c) single-frequency output obtained with the intracavity etalon.

6.3 SINGLE-WAVELENGTH OPERATION OF LASERS

Diode lasers can be made to operate at a single wavelength fairly easily, since the number of longitudinal modes decreases as the injection current is increased. Above a critical value of the injection current, oscillation in a single longitudinal mode is obtained.

Gas lasers require additional precautions to ensure operation in a single longitudinal mode. With He–Ne lasers, a simple solution is to use a short cavity, so that only one mode falls within the gain profile. A drawback of this technique is that the output power available is very low, typically around 0.1 mW. With higher-power lasers, it is necessary to use a mode selector (a short Fabry–Perot etalon) in the cavity. As shown in Figure 6.2(c), only one mode, that corresponding to the transmission peak of the etalon, then has a high enough gain to oscillate.

6.4 POLARIZATION OF LASER BEAMS

In a laser in which the plasma tube has windows sealed on at the Brewster angle (Brewster windows), reflection losses at the windows are least for waves polarized with the electric vector in the plane of incidence (see Appendix D). As a result, the laser generates a beam polarized in this plane.

In He–Ne lasers that use sealed-on end mirrors without Brewster windows and are oscillating in two or more longitudinal modes, mode competition results in alternate modes being orthogonally polarized. It is then worthwhile introducing a polarizer in the output beam to isolate a single polarization. This simple precaution can significantly improve the visibility of the fringes in an interferometer.

6.5 WAVELENGTH STABILIZATION OF LASERS

The wavelength of a free-running laser, even when it is operating in a single longitudinal mode, is not perfectly stable since it depends on the length of the optical path between the mirrors. With a diode laser, it is essential to stabilize the temperature of the laser with a thermoelectric cooler to minimize wavelength shifts due to temperature changes. With an He–Ne laser, stable operation can be obtained after an initial warm-up period of a few minutes, but residual wavelength variations of a few parts in 10^7 can be produced by mechanical vibrations and thermal effects. While such variations are acceptable for routine measurements, some method of wavelength stabilization is necessary where measurements are to be made with large optical path differences or with the highest precision.

The most common method of wavelength stabilization used with an He–Ne laser is by locking the output wavelength to the center of the gain curve. Techniques commonly used for this purpose include polarization stabilization, transverse Zeeman stabilization, and longitudinal Zeeman stabilization. Diode lasers can be stabilized by locking the output wavelength to a transmission peak of a temperature-controlled Fabry–Perot interferometer. Any of these methods can hold wavelength variations to less than 1 part in 10^8 over long periods.

6.6 LASER-BEAM EXPANSION

The beam from a laser oscillating in the TEM_{00} mode typically has a diameter ranging from a fraction of a millimetre to a few millimetres and a Gaussian intensity profile given by the relation

$$I(r) = \exp(-2r^2/w_0^2), \tag{6.3}$$

where r is the radial distance from the center of the beam. At a radial distance $r = w_0$, the intensity drops to $1/e^2$ of that at the center of the beam. Such a beam

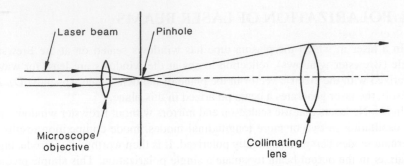

Figure 6.3. Arrangement used for expanding and spatially filtering a laser beam.

retains its Gaussian profile as it propagates, but its effective diameter increases due to diffraction. After traversing a distance z from the beam waist (the point at which the diameter of the beam is a minimum), the intensity distribution is given by the same relation with w_0 replaced by $w(z)$, where

$$w(z) = w_0\left[1 + \left(\lambda z / \pi w_0^2\right)^2\right]^{1/2}. \tag{6.4}$$

It can also be shown that, at a large distance from the beam waist, the angle of divergence of the beam is

$$\theta = \lambda / \pi w_0. \tag{6.5}$$

Many interferometers require a collimated beam filling a much larger aperture. In such a case, the laser beam is brought to a focus with a microscope objective; a lens with a suitable aperture can then be used, as shown in Figure 6.3, to obtain a collimated beam.

Due to the high degree of coherence of laser light, the expanded beam commonly exhibits random diffraction patterns (see Appendix O) produced by scratches or dust on the optics. These effects can be minimized by placing a pinhole (spatial filter) at the focus of the microscope objective. If the aperture of the microscope objective is greater than $2w_0$, the diameter of the focal spot is

$$d = 2\lambda f / \pi w_0, \tag{6.6}$$

where f is the focal length of the microscope objective; however, if only the central part of the laser beam (diameter D) is transmitted, the diameter of the focal spot is given by the relation

$$d = 2.44\lambda f / D. \tag{6.7}$$

If the pinhole is slightly smaller than the focal spot, randomly diffracted light is blocked, and the transmitted beam has a smooth profile.

6.7 PROBLEMS WITH LASER SOURCES

The high degree of coherence of laser light can result in some practical problems. One of these, as mentioned earlier, is random diffraction patterns (spatial noise) due to scattered light. When an extended source can be used, a simple solution is to introduce a rotating ground glass in the laser beam to wash out the speckle.

Another problem is the formation of spurious interference fringes due to stray light. Because light reflected or scattered from various surfaces in the optical path is coherent with the main beam, its amplitude a_s adds vectorially to the amplitude a of the main beam, as shown in Figure 6.4, resulting in a phase error

$$\Delta\phi = (a_s/a)\sin\phi, \tag{6.8}$$

where $\phi = (2\pi/\lambda)p$, p being the additional optical path traversed by the stray light. Since the amplitude of a light wave is proportional to the square root of its intensity, stray light with an intensity of only a few percent of the main beams can cause significant errors. To minimize these problems, a wedged beam splitter should be used, and stops (pinholes) must be introduced at suitable points in the optical path to cut out unwanted reflections and scattered light.

Yet another problem with a laser source is optical feedback. Light reflected back to the laser can cause changes in its power output and even its frequency. Optical feedback can be minimized by a combination of a polarizer and a $\lambda/4$ plate (an optical isolator) which rotates the plane of polarization of the return beam by 90°. Needless to say, both the polarizer and the $\lambda/4$ plate should be tilted slightly to eliminate back reflections.

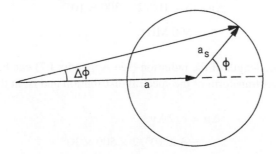

Figure 6.4. Phase error produced by scattered laser light.

6.8 LASER SAFETY

The beam from a laser is focused by the lens of the eye to a very small spot on the retina. As a result, the direct beam from even a low-power He–Ne laser (0.5 mW) can cause serious eye damage. Reflections off shiny surfaces can also be dangerous. Carbon dioxide lasers are particularly dangerous, since the beam cannot be seen, and their output power is much higher. The danger is much less once the beam is expanded. However, when working with lasers, always TAKE CARE TO PROTECT YOUR EYES (see *ANSI Standard Z* 136.1-1993).

6.9 SUMMARY

- Laser sources eliminate most of the problems of conventional thermal sources.
- Interference with large optical path differences requires operation in a single longitudinal mode.
- Wavelength stabilization is advisable for accurate measurements.
- Eliminate scattered light and back reflections.
- TAKE CARE OF YOUR EYES.

6.10 PROBLEMS

Problem 6.1. *An He–Ne laser with a 300-mm-long resonant cavity oscillates in two longitudinal modes. What is the frequency difference between these modes? What is the coherence length of the radiation?*

From Eq. 6.2, the frequency difference between adjacent modes is

$$\Delta v = 3 \times 10^8 / 2 \times 300 \times 10^{-3}$$
$$= 500 \, \text{MHz}. \tag{6.9}$$

The coherence length of the radiation (see Appendix I.7) can be evaluated by taking the Fourier transform of the spectral distribution and is, in this case,

$$\Delta p = c / 2 \Delta v$$
$$= 3 \times 10^8 / 2 \times 500 \times 10^6$$
$$= 0.3 \, \text{m}. \tag{6.10}$$

Problem 6.2. *If the Doppler-broadened gain profile for an He–Ne laser ($\lambda = 633$ nm) has a width of 1.4 GHz, what is the maximum length that the laser cavity can have to ensure operation in a single longitudinal mode?*

If the laser is to operate in a single longitudinal mode, the separation of adjacent cavity resonances must be greater than the width of the gain profile. From Eq. 6.2,

$$L_{\max} = c/2\Delta\nu$$
$$= 3 \times 10^8/2 \times 1.4 \times 10^9$$
$$= 107 \text{ mm.} \tag{6.11}$$

Problem 6.3. *In the arrangement shown in Figure 6.3, the central part of the beam from an He–Ne laser ($\lambda = 633$ nm) is isolated by an aperture with a diameter of 2.0 mm and brought to a focus by a microscope objective with a focal length of 32 mm. What would be a suitable size for the pinhole?*

From Eq. 6.7, the diameter of the focal spot is

$$d = 2.44 \times 0.633 \times 10^{-6} \times 32 \times 10^{-3}/2 \times 10^{-3}$$
$$= 24.4 \ \mu\text{m.} \tag{6.12}$$

A pinhole with a diameter of 20 μm would ensure a clean beam, with only a marginal loss of light.

Problem 6.4. *The beam from a 0.5 mW He–Ne laser ($\lambda = 633$ nm, $w_0 = 1.0$ mm) is brought to a focus on the retina by the lens of the human eye (effective focal length 25 mm). What is the power density in the focused spot?*

From Eq. 6.6, the diameter of the focal spot is

$$d = 2 \times 0.633 \times 10^{-6} \times 25 \times 10^{-3}/\pi \times 10^{-3}$$
$$= 10.1 \ \mu\text{m.} \tag{6.13}$$

The power density in the focal spot is, therefore,

$$I = 0.5 \times 10^{-3}/\pi \times 5.05^2 \times 10^{-12}$$
$$= 6.2 \times 10^6 \text{ W/m}^2, \tag{6.14}$$

which is about 100 times that produced by looking directly at the sun.

FURTHER READING

For more information, see

1. D. C. O'Shea, W. R. Callen, and W. T. Rhodes, *An Introduction to Lasers and Their Applications*, Addison-Wesley, Reading, MA (1977).
2. O. Svelto, *Principles of Lasers*, Plenum Press, New York (1989).
3. W. T. Silvfast, *Laser Fundamentals*, Cambridge University Press, Cambridge, UK (1996).
4. R. Henderson and K. Schulmeister, *Laser Safety*, Institute of Physics, Bristol (2004).

7

Photodetectors

In this chapter we discuss some types of photodetectors that are commonly used with interferometers. They include

- Photomultipliers
- Photodiodes
- Charge-coupled detector arrays
- Photoconductive detectors
- Pyroelectric detectors

7.1 PHOTOMULTIPLIERS

In a photomultiplier, light is incident on a photocathode in an evacuated glass envelope. As shown in Figure 7.1, the same envelope also contains a set of electrodes, called *dynodes*, located between the cathode and the anode, which are held at successively higher potentials. The electrons emitted by the cathode are electrostatically focused on the first dynode, where they produce a much larger number of secondary electrons; this process continues until the electrons from the last dynode reach the anode. Typically, the potential drop across the tube ranges from a few hundred volts to a few kilovolts, and the overall gain may be as high as 10^{11}. The frequency response of a photomultiplier is limited mainly by the spread in the transit times of the secondary electrons, which can be kept as low as 10 ns by proper design. Photomultipliers have extremely high sensitivity in the UV and visible regions, but their sensitivity falls off rapidly in the near infrared.

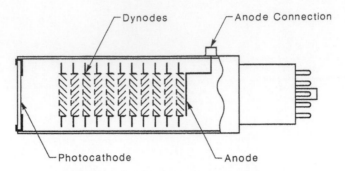

Figure 7.1. Construction of a photomultiplier.

7.2 PHOTODIODES

Photodiodes use a junction between p- and n-type semiconductors to detect light. An n-type semiconductor contains many highly mobile electrons, while a p-type material contains less mobile positive holes. When two such materials are joined, the electrons and holes are drawn to opposite sides of the junction, and an energy level structure similar to that shown in Figure 7.2 is obtained. The region near the junction contains virtually no electrons or holes and is known as the depletion layer.

When the junction is illuminated, valence-band electrons are excited to the conduction band, creating electron-hole pairs. Because of the strong potential gradient in the junction region, the electrons and holes are accelerated in opposite directions, and a current flows.

The speed of response and sensitivity of a photodiode can be increased by reverse biasing; the positive side of a battery is connected to the n-type material and

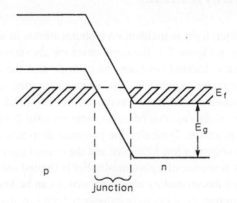

Figure 7.2. Energy level structure of a p-n junction.

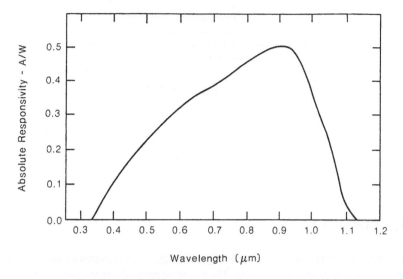

Figure 7.3. Typical spectral response of a silicon photodiode.

the negative side to the p-type material. Higher sensitivity can also be obtained by introducing a layer of a high-resistivity (intrinsic) material between the p- and n-layers; such a device is known as a p-i-n (or PIN) diode. PIN diodes have a useful response up to a frequency of a few hundred MHz.

With a sufficiently high reverse bias, electron multiplication due to secondary emission can occur. This effect is utilized in avalanche photodiodes to obtain a gain in sensitivity by a factor of a few hundred, but at the expense of an increase in noise at low light levels. Photodiodes are also available in a package that contains a high-gain operational amplifier. These devices can be used at very low light levels and, unlike photomultipliers, require only a low voltage. A linear relationship between output voltage (or current) and the light level can be obtained over several decades.

Silicon photodiodes are the most commonly used and, as shown in Figure 7.3, have a peak sensitivity around 0.8–0.9 μm. Germanium and InGaAs photodiodes are useful in the region from 1.1 to 1.7 μm.

7.3 CHARGE-COUPLED DETECTOR ARRAYS

Charge-coupled detector (CCD) arrays have made possible simultaneous measurements of light intensities at a number of points and have opened up many new possibilities in interferometry.

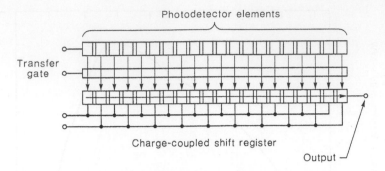

Figure 7.4. Schematic of a linear CCD sensor.

7.3.1 Linear CCD Sensors

A linear CCD sensor consists of a linear array of photosensors and an associated, charge-coupled shift register. These are separated, as shown in Figure 7.4, by an electrode known as a transfer gate. In operation, the charges collected by the individual photosensor elements over a fixed integration time are transferred to the corresponding elements of the shift register. This charge pattern is then moved along the shift register and read out during the next integration period.

7.3.2 Area CCD Sensors

In an area CCD sensor, as shown in Figure 7.5, the charges accumulating in the photosensor elements in each column are transferred, at the end of each integration period, to the adjacent, vertical shift registers. The contents of each of the vertical shift registers are then transferred, one charge packet at a time, to the horizontal shift register at the top of the array. Each transfer from the vertical registers fills the horizontal register, which is then read out to produce a line of a video signal. After all the vertical shift registers have been read out to produce a complete video frame, the process begins again.

7.3.3 Frame-Transfer CCD Sensors

In frame-transfer CCD sensors the array is divided into two identical areas, as shown in Figure 7.6. One area is the image zone, which is illuminated, and the other is a masked memory zone, into which the image information is transferred for subsequent readout.

In this architecture, each column of photosensitive elements in the image zone constitutes a CCD shift register, separated from the others by an insulating wall. At the end of each integration period, during the field blanking period (<1 ms),

Charge-coupled shift register

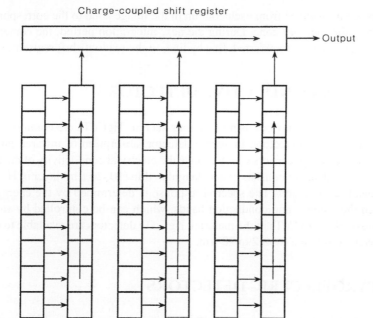

→ Output

— Charge-coupled shift register

— Photodetector elements

Figure 7.5. Schematic of an area CCD sensor.

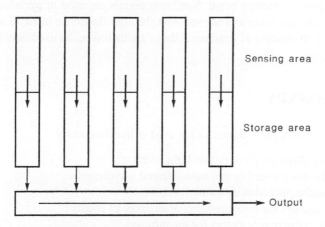

Sensing area

Storage area

→ Output

Charge-coupled shift register

Figure 7.6. CCD sensor using frame transfer.

charges are transferred from each column in the image zone to the corresponding column in the memory zone. During the next integration period, the contents of the memory zone are transferred, line by line, to the readout shift register.

7.4 PHOTOCONDUCTIVE DETECTORS

Photoconducting devices using materials such as HgCdTe are commonly employed as infrared detectors. In such a detector, absorption of infrared photons produces free charge carriers which change the electrical conductivity of the material. A typical detector consists of a rectangular, thin ($10\text{--}20\ \mu$m) layer of HgCdTe with metalized contacts. The spectral response is determined by the energy gap between the valence and conduction bands, which can be controlled by varying the ratio of HgTe to CdTe in the material. HgCdTe detectors are available to cover the wavelength range from 2 to 20 μm.

7.5 PYROELECTRIC DETECTORS

Pyroelectric detectors use a ferroelectric material, such as lead zirconate ceramic, or a plastic, such as polyvinylidene fluoride, that is electrically polarized by cooling it from an appropriate temperature in an electrical field. If the material is then placed between two electrodes, any change in its temperature, due to absorption of infrared radiation, produces a current in the external circuit. In a pyroelectric vidicon, the charge distribution on one face of a plastic film is read out by a scanning electron beam. Similar materials are used in pyroelectric CCD arrays. Pyroelectric detectors are sensitive through the entire infrared region, but respond only to changes of irradiance; they can, therefore, be used only with modulated sources.

7.6 SUMMARY

Several types of photodetectors are used in interferometry:

- Photomultipliers for very low light levels
- PIN diodes for visible and near-infrared wavelengths
- Avalanche photodiodes for high sensitivity
- CCD sensors for measurements at an array of points
- Photoconductive detectors for the infrared
- Pyroelectric detectors for the far infrared
- Pyroelectric detectors require a modulated source

7.7 PROBLEMS

Problem 7.1. *What type of photodetector would you use for a fringe-counting interferometer with a laser-diode source?*

A silicon photodiode is almost ideal for this application, since it has a small sensitive area, and its peak sensitivity matches the laser wavelength. In addition, its frequency response is more than adequate and it only requires a low-voltage power supply.

Problem 7.2. *The collimating lens in a Fizeau interferometer (see Section 9.1) has an aperture of 100 mm and a focal length of 1000 mm. The photodetector is a 488 × 380 element CCD array with an 8.8 × 11.4 mm active area. A second lens is to be used to image the interference fringes on the photodetector. What would be a suitable focal length for this lens?*

The imaging lens has to be placed so that its front focal plane is located at the back focal plane of the collimating lens. The diameter of the image of the pupil of the collimating lens formed by the imaging lens must then be less than 8.8 mm for it to fit within the dimensions of the array. Since the pupil of the collimating lens has a diameter of 100 mm, the imaging lens should have (to a first approximation) a maximum focal length

$$f = 1000 \times 8.8/100$$

$$= 88 \text{ mm.} \tag{7.1}$$

Problem 7.3. *A microscope with a 16-mm objective (0.2 NA) is used to view the interference fringes produced between a reference flat surface and a diamond-turned surface. A linear CCD array containing 1728 elements, with a center-to-center spacing of 13 μm, is positioned at the primary image formed 160 mm behind the microscope objective. Is the spacing of the detector elements close enough to avoid significant loss of image resolution?*

The lateral resolution in the object plane is

$$\Delta x_0 = 1.22 \times 0.633 \times 10^{-6}/0.2$$

$$= 3.86 \ \mu\text{m,} \tag{7.2}$$

which would correspond to a distance in the image plane,

$$\Delta x_1 = 3.86 \times 160/16$$

$$= 38.6 \ \mu\text{m.} \tag{7.3}$$

Since the spacing of the elements on the photodetector is around one third of this distance, there should not be a significant loss of resolution.

FURTHER READING

For more information, see

1. R. W. Boyd, *Radiometry and the Detection of Optical Radiation*, John Wiley, New York (1983).
2. D. F. Barbe, *Charge-Coupled Devices*, Topics in Applied Physics, Vol. 38, Springer-Verlag, Berlin (1980).
3. T. J. Tredwell, *Visible Array Detectors*, Chap. 22 in Handbook of Optics, Vol. I, Ed. M. Bass, McGraw-Hill, New York (1995).
4. L. J. Kozlowski and W. F. Kosonocky, *Infrared Detector Arrays*, Chap. 23 in Handbook of Optics, Vol. I, Ed. M. Bass, McGraw-Hill, New York (1995).
5. G. C. Holst, *CCD Arrays, Cameras and Displays*, Vol. PM 57, SPIE Press, Bellingham, WA (1998).

8

Measurements of Length

An important application of interferometry is in accurate measurements of length. In this chapter, we discuss

- The definition of the metre
- Length measurements
- Measurements of changes in length

8.1 THE DEFINITION OF THE METRE

A problem with the standard metre bar was that measurements could only be repeated to a few parts in 10^7. Michelson was the first to show that an improvement by an order of magnitude was possible with interferometric measurements using the red cadmium line. After an extensive search for a suitable spectral line, the standard metre bar was finally abandoned in 1960, and the metre was redefined in terms of the wavelength of the orange line from a ^{86}Kr discharge lamp. However, when frequency stabilized lasers became available, comparisons of their wavelengths with the ^{86}Kr standard showed that the accuracy of such measurements was limited to a few parts in 10^9 by the uncertainties associated with the ^{86}Kr standard. This led to a renewed search for a better definition of the metre.

The primary standard of time is the ^{133}Cs clock. Since laser frequencies could be compared with the ^{133}Cs clock frequency with an accuracy of a few parts in 10^{11}, the metre was redefined in 1983 as follows:

The metre is the length of the path traveled by light in vacuum during a time interval of $1/299\,792\,458$ *of a second.*

The speed of light (a basic physical constant) is now fixed, and length becomes a quantity derived from measurement of time or its reciprocal, frequency. Practical measurements of length are carried out by interferometry, using the vacuum wavelengths of stabilized lasers whose frequencies have been compared with the ^{133}Cs standard. Several lasers are now available, whose wavelengths have been measured extremely accurately. With such lasers, optical interferometry can be used for very accurate measurements of distances up to a hundred metres.

8.2 LENGTH MEASUREMENTS

Measurements of the lengths of end standards (gauge blocks) can be made with a Kosters interferometer. As shown in Figure 8.1, this is a Michelson interferometer using collimated light and a dispersing prism to select any single spectral line from the source. The end standard is contacted to one of the mirrors of the interferometer (a polished, flat, metal surface) so that interference fringes are obtained, as shown in Figure 8.2, between the free end of the end standard and the reference mirror, as well as between the two mirrors.

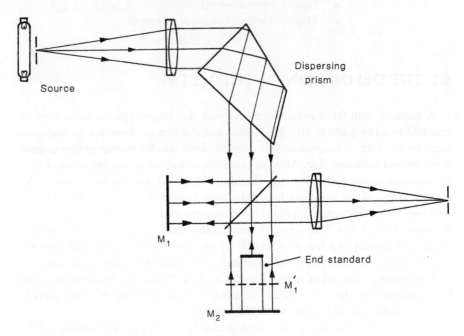

Figure 8.1.   Kosters interferometer for end standards.

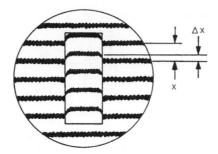

Figure 8.2. Interference fringes in a Kosters interferometer.

8.2.1 The Fractional-Fringe Method

To measure the length of an end standard, we have to evaluate the difference between the interference orders for the surrounding field and the free end of the end standard. However, if the difference between the interference orders is, say, $(m + \epsilon)$, where m is an integer and ϵ is a fraction, the interference fringes only give (see Figure 8.2) the fractional part $\epsilon = \Delta x / x$. If the length of the end standard is known within a few micrometres, a simple method of obtaining the integral part m is from observations of the fractional part ϵ with two or more wavelengths. A series of values for m that cover this range of lengths are then set up for one wavelength, and the corresponding calculated values of the fractional part ϵ for the other wavelengths are compared with the observed values. The value of m that gives the best fit at all the wavelengths is then chosen.

8.2.2 Fringe Counting

A more direct method of measuring lengths is to count the fringes that pass a given point in the field while one mirror of the interferometer is moved over the distance to be measured. This can now be done quickly and easily with photoelectric fringe-counting techniques. An optical system is used to produce two interferograms that yield output signals that are in phase quadrature. These signals can be processed to correct for vibration or retraced motion.

8.2.3 Heterodyne Techniques

Interferometers using heterodyne techniques are now widely employed for length measurements. In the Hewlett-Packard interferometer, shown schematically in Figure 8.3, an He–Ne laser is forced to oscillate simultaneously at two frequencies, separated by a constant difference of about 2 MHz, by applying an axial magnetic field. These two waves, which are circularly polarized in opposite

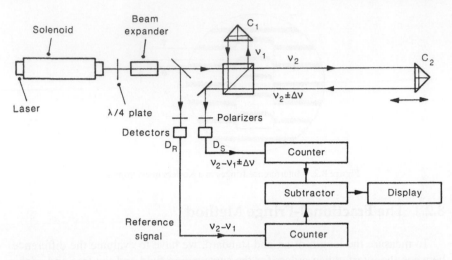

Figure 8.3. Fringe-counting interferometer (after J. N. Dukes and G. B. Gordon, *Hewlett-Packard Journal* **21**, No. 12, 2–8, Dec. 1970). ©1970 by Hewlett-Packard Company. Reproduced with permission.

senses, are converted to orthogonal linear polarizations by a $\lambda/4$ plate. A polarizing beam splitter reflects one wave to a fixed corner reflector C_1, while the other is transmitted to a movable corner reflector C_2. Both waves return along a common axis and pass through a polarizer that brings them into a condition to interfere.

The signals at the difference frequency (see Appendix J), from the detector D_S and a reference detector D_R, go to a differential counter. If the two reflectors are stationary, the frequencies of the two signals are the same, and no net count accumulates. If one of the reflectors is moved, the change in the optical path, in wavelengths, is given by the net count.

The Hewlett-Packard interferometer is now used widely in industry for measurements over distances up to 60 m. With additional optics, it can also be used for measurements of angles, straightness, flatness, and squareness.

An alternative method of producing a two-frequency laser beam is to use an acousto-optic frequency shifter (see Appendix K). This method has the advantage that the frequency difference can be much higher, so that higher count rates can be handled.

8.2.4 Synthetic Long-Wavelength Signals

Another technique, which can be used if the distance to be measured is known approximately, involves synthetic long-wavelength signals. This technique is based on the fact that if a two-beam interferometer is illuminated simultaneously with two wavelengths, λ_1 and λ_2, the envelope of the fringes corresponds to the interference pattern that would be obtained with a much longer synthetic wavelength

$$\lambda_s = \lambda_1 \lambda_2 / |\lambda_1 - \lambda_2|. \tag{8.1}$$

The carbon dioxide laser can operate at several closely spaced wavelengths that have been measured accurately and is, therefore, well suited to such measurements. The laser is switched rapidly between two of these wavelengths, and the output signal obtained from a photodetector, as one of the interferometer mirrors is moved, is squared, low-pass filtered, and processed in a computer to obtain the phase difference. Distances up to 100 m can be measured with an accuracy of 1 part in 10^7.

8.2.5 Frequency Scanning

Yet another method of measuring lengths is to use a diode laser whose frequency is swept linearly with time by controlling the injection current. In the arrangement shown in Figure 8.4, interference takes place between the beams reflected from the front surface of a fixed reflector and a movable reflector. An optical path difference p introduces a time delay p/c between the two beams, where c is the speed of light, and the beams interfere at the detector to yield a beat signal with a frequency

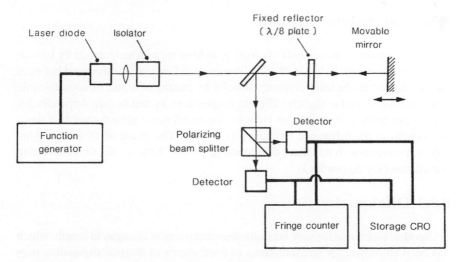

Figure 8.4. Interferometer using laser frequency scanning for measurements of distances (T. Kubota, M. Nara, and T. Yoshino, *Opt. Lett.* **12**, 310–312, 1987).

$$\Delta v = (p/c)(dv/dt), \tag{8.2}$$

where dv/dt is the rate at which the laser frequency is varying with time.

8.2.6 Environmental Effects

All such measurements must be corrected for the actual value of the refractive index of air, which depends on the temperature and the relative humidity. In addition, care must be taken to minimize the initial difference between the optical paths, in air, in the two arms of the interferometer (the dead path), to reduce errors due to changes in the environmental conditions.

8.3 MEASUREMENTS OF CHANGES IN LENGTH

8.3.1 Phase Compensation

Very accurate measurements of changes in length are possible by methods based on phase compensation. Changes in the output intensity from the interferometer are detected and fed back to a phase modulator (a piezoelectric translator on which one of the mirrors is mounted) so as to hold the output constant. The drive signal to the modulator is then a measure of the changes in the length of the optical path.

8.3.2 Heterodyne Methods

Very accurate measurements of changes in length can also be made by heterodyne interferometry. In one method, a frequency difference is introduced between the two beams in the interferometer, usually by means of a pair of acousto-optic modulators operated at slightly different frequencies, v_1 and v_2 (see Appendix K). The output from a photodetector viewing the interference pattern then contains a component at the difference frequency $(v_1 - v_2)$. The phase of this heterodyne signal corresponds to the phase difference $(\phi_1 - \phi_2)$ between the two interfering wavefronts (see Appendix J).

8.3.3 Dilatometry

Another technique for very accurate measurements of changes in length, which has been employed for measurements of coefficients of thermal expansion, uses the heterodyne signal produced by superposing the beams from two lasers operating on the same spectral transition.

For this purpose, two partially transmitting mirrors are attached to the ends of the specimen, forming a Fabry–Perot interferometer. A servo system is used to lock the output wavelength of one of the lasers to a transmission peak of this Fabry–Perot interferometer. The wavelength of this slave laser is then an integral submultiple of the optical path difference in the interferometer. A displacement of one of the mirrors results in a change in the wavelength of the slave laser and, hence, in its frequency. These changes are measured by mixing the beam from the slave laser with the beam from the other laser (a frequency-stabilized reference laser) at a fast photodiode and measuring the frequency of the beats.

8.4 SUMMARY

- Length is now a quantity defined in terms of the speed of light (which is fixed) and time (or its reciprocal, frequency).
- Measurements of length are made with lasers whose frequencies (wavelengths) have been measured accurately.
- Measurements of length can be made by:
  - measurements at two or more wavelengths
  - electronic fringe counting
  - heterodyne techniques
  - laser frequency scanning.
- Corrections must be made for the refractive index of air.
- Measurements of changes in length can be made by phase compensation or by heterodyne methods.

8.5 PROBLEMS

Problem 8.1. *The following values are obtained for the fractional fringe order in a Kosters interferometer with an end standard, using the red, green, and blue spectral lines from a low-pressure cadmium lamp.*

Wavelength (nm)	Measured fraction
643.850	0.1
508.585	0.0
479.994	0.5

Mechanical measurements have established that the length of the end standard is 10 ± 0.001 mm. What is its exact length?

Since the length of the end standard is between 10.001 and 9.999 mm, the value of the integral interference order N for the red line ($\lambda = 643.850$ nm) must lie between 31,060 and 31,066. The measured value of the fractional interference order for this line is 0.1. Accordingly, we take values of the interference order ranging from 31,060.1 to 31,066.1 for the red line and calculate the corresponding values of the length, as well as the interference orders for the other lines, as shown in the following table:

Measured fractions			Length (mm)
Red	Green	Blue	
31,060.1	39,321.0	41,663.2	9.99903
31,061.1	39,322.3	41,664.5	9.99935
31,062.1	39,323.5	41,665.9	9.99967
31,063.1	39,324.8	41,667.2	10.00000
31,064.1	39,326.0	41,668.5	10.00031
31,065.1	39,327.3	41,669.9	10.00064
31,066.1	39,328.6	41,671.2	10.00096

From these figures, we see that the only value for the length of the end standard that produces satisfactory agreement between the measured and calculated values of the fractional interference order for the green and blue lines is 10.0003 mm.

Problem 8.2. *The wavelengths (in air) of three spectral lines from a CO_2 laser are $\lambda_1 = 10.608565$ μm, $\lambda_2 = 10.271706$ μm, and $\lambda_3 = 10.257656$ μm. What are the synthetic wavelengths that can be produced?*

Three synthetic wavelengths can be generated using pairs of these lines. From Eq. 8.1, the values of these synthetic wavelengths are

Wavelengths used	Synthetic wavelength
λ_1 and λ_3	310.1 μm
λ_1 and λ_2	323.5 μm
λ_2 and λ_3	7.499 mm

Problem 8.3. *In an interferometer using a diode laser as the light source, the injection current of the laser is modulated at a frequency of 90 Hz by a triangular wave with a peak-to-peak amplitude of 15.0 mA. The frequency of the laser changes with the injection current at a rate $(d\nu/dI) = 4.1$ GHz/mA. If a beat signal with a frequency of 3.690 kHz is obtained, what is the optical path difference in the interferometer?*

Since the rate of change of the injection current with time is

$$(dI/dt) = 15 \times 180 = 2700 \text{ mA/sec}, \tag{8.3}$$

the rate of change of the laser frequency with time is

$$(d\nu/dt) = (d\nu/dI)/(dI/dt)$$
$$= 4.1 \times 2700$$
$$= 11{,}070 \text{ GHz/sec}. \tag{8.4}$$

Accordingly, from Eq. 8.2, the optical path difference in the interferometer is

$$p = c\Delta\nu(d\nu/dt)$$
$$= 2.998 \times 10^8 \times 3690/11.070 \times 10^{12}$$
$$= 0.999 \text{ m}. \tag{8.5}$$

Problem 8.4. *A Fabry–Perot interferometer (FPI) made up of two mirrors attached to a 100-mm-long fused silica tube is set up in an evacuated oven. The output from an He–Ne laser ($\lambda = 632.8$ nm), which is locked to a transmission peak of the FPI, is mixed with the output from a frequency-stabilized reference laser at a fast photodiode, and the frequency of the resulting beat is measured. A change of $1.0°C$ in the temperature of the oven is found to produce a change in the beat frequency of 235.5 MHz. What is the coefficient of thermal expansion of the silica tube?*

From Eq. 6.1, we can express the relationship between ΔL, the change in spacing of the mirrors of the FPI, and $\Delta\nu$, the corresponding change in the frequency of the transmission peak, in the form

$$\Delta\nu/\nu = -\Delta L/L. \tag{8.6}$$

Since the nominal frequency of the He–Ne laser is

$$\nu = c/\lambda = 4.738 \times 10^{14} \text{ Hz}, \tag{8.7}$$

the coefficient of thermal expansion of the silica tube is

$$\Delta L/L = 235.5 \times 10^6/4.738 \times 10^{14}$$
$$= 0.497 \times 10^{-6}/°C. \tag{8.8}$$

FURTHER READING

For more information, see

1. P. Hariharan, *Interferometry with Lasers*, in Progress in Optics, Vol. XXIV, Ed. E. Wolf, North-Holland, Amsterdam (1987), pp. 103–164.
2. P. Hariharan, *Interferometric Metrology: Current Trends and Future Prospects*, Proc. SPIE, Vol. 816, 2–18 (1988).

9

Optical Testing

Another major application of interferometry is in testing optical components and optical systems.

Some of the topics that we will discuss in this chapter are

- The Fizeau interferometer
- The Twyman–Green interferometer
- Analysis of wavefront aberrations
- Laser unequal-path interferometers
- The point-diffraction interferometer
- Shearing interferometers
- Grazing-incidence interferometers

9.1 THE FIZEAU INTERFEROMETER

A polished flat surface can be compared with a standard reference flat surface by putting them together and viewing the interference fringes (fringes of equal thickness) formed in the thin air film separating them. A light box, such as that shown in Figure 9.1, with a sodium vapor or mercury vapor lamp as the source, can be used for this purpose. The fringe pattern corresponds to a contour map of the errors of the test surface. A simple method to test whether the test surface is convex or concave is to apply gentle pressure at a point near its edge. If the surface is convex, the center of the fringe pattern moves toward this point; if the surface is concave, the center of the fringe pattern moves away from it.

To measure surface errors smaller than a wavelength, one of the plates is tilted slightly to produce a wedged air film and introduce a few fringes across the field.

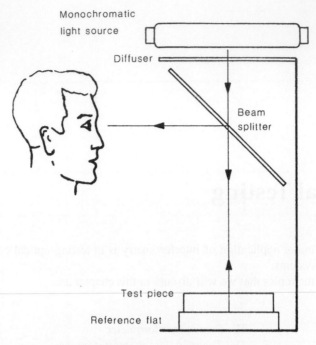

Figure 9.1. Light box used for viewing interference fringes of equal thickness formed between two flat surfaces.

The shape of the diametrical fringe then indicates the deviations of the surface from a plane. If the average fringe spacing is x, and the distance between two parallel straight lines enclosing the diametrical fringe is Δx, the peak-to-valley error (P-V error) of the surface is $(\Delta x / x)(\lambda / 2)$. If a standard reference flat is not available, a set of three nominally flat surfaces, A, B, and C, can be tested in pairs $(A + B, B + C, C + A)$ to evaluate their individual deviations from flatness along a diameter.

Because of the risk of damage to the test and reference surfaces, it is desirable not to place them in contact, but to have them separated by a small air gap; it is then necessary to use collimated light. A typical setup for this purpose (the Fizeau interferometer) is shown in Figure 9.2.

Other applications of the Fizeau interferometer include checking the faces of a transparent plate for parallelism and checking slabs of optical glass for homogeneity. Concave and convex surfaces can also be tested against a reference flat surface by using a well-corrected converging lens, as shown in Figure 9.3. A laser operating in a single longitudinal mode (see Section 6.3) must be used as the source to obtain interference fringes with good visibility at such large optical path differences.

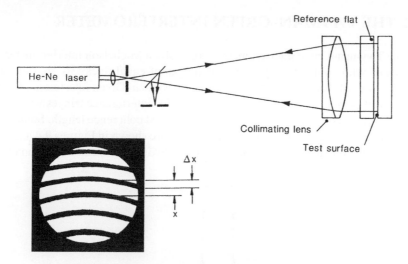

Figure 9.2. Fizeau interferometer for testing flat surfaces.

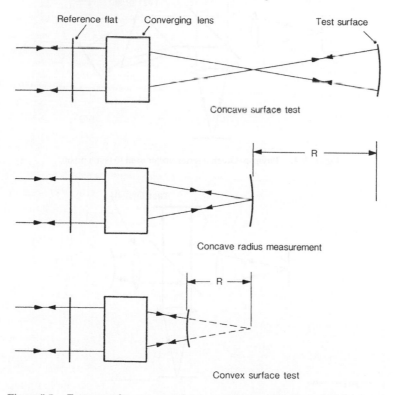

Figure 9.3. Test setups for concave and convex surfaces using a Fizeau interferometer.

9.2 THE TWYMAN–GREEN INTERFEROMETER

The Twyman–Green interferometer is basically a Michelson interferometer illuminated with collimated light, so that fringes of equal thickness are obtained (see Section 3.4.3 and Appendix F). With the Twyman–Green interferometer, the two optical paths can be made nearly equal, so that interference fringes with good visibility can be obtained with light having a limited coherence length. Some typical optical setups for tests on prisms and lenses are shown in Figures 9.4 and 9.5. Similar test geometries can also be implemented with the Fizeau interferometer.

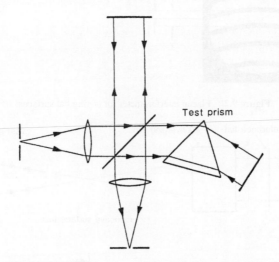

Figure 9.4. Twyman–Green interferometer used to test a prism.

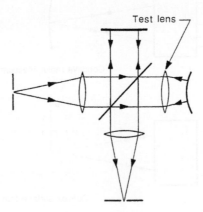

Figure 9.5. Twyman–Green interferometer used to test a lens.

Figure 9.6. Twyman–Green interferograms for some typical lens aberrations: (left to right) tilt, defocusing, astigmatism, coma, and spherical aberration.

9.3 ANALYSIS OF WAVEFRONT ABERRATIONS

Both the Fizeau interferometer and the Twyman–Green interferometer are commonly used for tests on complete optical systems. It is often necessary, then, to analyze the interferogram to determine the nature and magnitude of the various aberrations that are present.

Some interferograms for typical lens aberrations are shown in Figure 9.6. They correspond (left to right) to tilt, defocusing, astigmatism, coma, and spherical aberration.

In Cartesian coordinates, the deviation of the test wavefront from a reference sphere centered on the image point can be written as

$$W(x, y) = \sum_{k=0}^{n} \sum_{l=0}^{k} c_{kl} x^l y^{k-l}. \tag{9.1}$$

If we consider only the primary aberrations, Eq. 9.1 becomes

$$W(x, y) = A\left(x^2 + y^2\right)^2 + By\left(x^2 + y^2\right) + C\left(x^2 + 3y^2\right) + D\left(x^2 + y^2\right) + Ey + Fx, \tag{9.2}$$

where A is the coefficient of spherical aberration, B is the coefficient of coma, C is the coefficient of astigmatism, D is the coefficient of defocusing, and E and F give the tilt about the x- and y-axes, respectively.

Alternatively, the wavefront deviations can be expressed in polar coordinates as a linear combination of Zernike circular polynomials in the form

$$W(\rho, \theta) = \sum_{k=0}^{n} \sum_{l=0}^{k} \rho^k (A_{kl} \cos l\theta + B_{kl} \sin l\theta), \tag{9.3}$$

where ρ and θ are polar coordinates over the pupil, and $(k - l)$ is an even number.

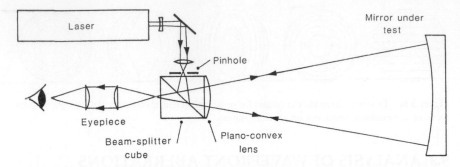

Figure 9.7. Laser unequal-path interferometer (R. V. Shack and G. W. Hopkins, *Opt. Eng.* **18**, 226–228, 1979).

In either case, the first step is to obtain, from the interferogram, the optical path differences at a suitably chosen array of points; the aberration coefficients can then be calculated from a set of linear equations.

For accurate measurements, it is necessary to use an auxiliary lens to image the surface of the element under test on the photodetector or the film used to record the interferogram. This precaution is essential if the test wavefront exhibits significant amounts of aberration.

9.4 LASER UNEQUAL-PATH INTERFEROMETERS

With a laser operating in a single longitudinal mode, interference fringes can be obtained even with large optical path differences. This has led to the development of compact optical systems based on the Twyman–Green and Fizeau interferometers that can be used for testing large optics.

Figure 9.7 shows a typical optical system that uses a beam-splitting cube with a plano-convex lens cemented to one surface (the Shack cube interferometer). The beam from a laser is brought to a focus at a pinhole placed at the image of the center of curvature of the convex surface formed in the beam splitter. Interference fringes are produced by the beams reflected from the test surface and the convex surface of the beam-splitting cube.

9.5 THE POINT-DIFFRACTION INTERFEROMETER

The point-diffraction interferometer consists of a small pinhole in a partially transmitting film ($T \approx 0.05$) placed at the focus of the converging test wavefront. As shown schematically in Figure 9.8, interference takes place between the test wavefront, which is transmitted by the film, and a spherical reference wave produced by diffraction at the pinhole. The fringe pattern is similar to those obtained

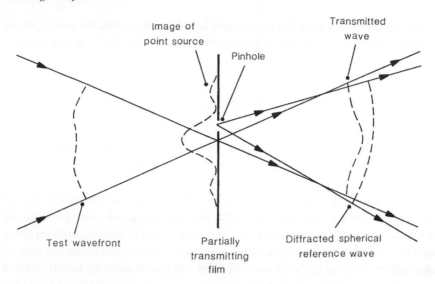

Figure 9.8. Point-diffraction interferometer (R. N. Smartt and W. H. Steel, *Japan J. Appl. Phys.* **14**, Suppl. 14-1, 351–356, 1975).

with the Fizeau and Twyman–Green interferometers and corresponds to a contour map of the wavefront aberrations. A nematic liquid-crystal layer can be used to introduce phase shifts (see Section 10.2) between the object beam and the reference beam generated by a microsphere embedded within the liquid-crystal layer.

The point diffraction interferometer has the advantages of simplicity and ease of use. It can be used, for instance, to test a telescope objective *in situ*, using a bright star as the light source. Its disadvantage is its low transmittance.

9.6 SHEARING INTERFEROMETERS

In a shearing interferometer, the interference pattern is produced by superposing two images of the test wavefront.

Shearing interferometers have the advantage that no reference surface is required, and a very simple and compact optical system can be used to test large surfaces. In addition, since both beams traverse very nearly the same optical path, the fringe pattern is less affected by air currents and vibration than in a conventional interferometer. However, shearing interferometers have the disadvantage that numerical analysis of the interferogram is necessary to obtain the wavefront errors. Moreover, since interference takes place between beams derived from different parts of the test wavefront, it is necessary to use a source, such as a laser, that produces light with a high degree of spatial coherence.

Many types of shearing interferometers have been described, but two types are commonly used: lateral shearing interferometers and radial shearing interferometers.

9.6.1 Lateral Shearing Interferometers

In a lateral shearing interferometer, two images of the test wavefront, of the same size, are superposed with a small mutual lateral displacement, as shown in Figure 9.9.

If the shear s is a small fraction of the diameter of the test wavefront, the optical path difference at any point in the interference pattern corresponds to the derivative of the wavefront errors (i.e., the slope errors of the test surface) along the direction of shear (see Appendix L.1). The wavefront aberrations can then be obtained by integrating the phase data from two lateral shearing interferograms with orthogonal directions of shear.

Typical lateral shearing interferograms for the primary aberrations are presented in Figure 9.10, along with the corresponding Twyman–Green interferograms.

A simple lateral shearing interferometer using two identical gratings ($\tilde{2}00$ lines/mm) is shown in Figure 9.11. The shear is varied by rotating one grating in its own plane. The spacing and orientation of the fringes are controlled by the separation of the gratings and their position relative to the focus.

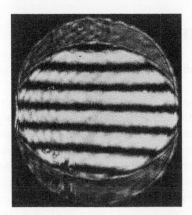 

Figure 9.9. Typical interferograms obtained with a lateral shearing interferometer.

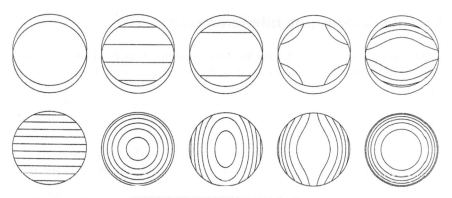

Figure 9.10. Interferograms obtained with a lateral shearing interferometer ($s = 0.2$) for (left to right) tilt, defocusing, astigmatism, coma, and spherical aberration. The corresponding interferograms obtained with a Twyman–Green interferometer are shown in the lower row.

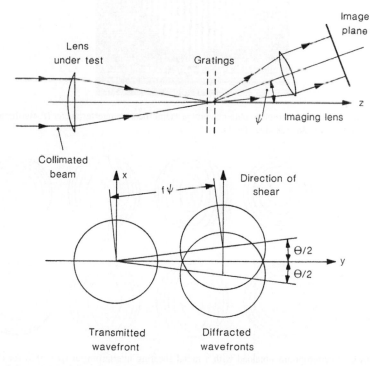

Figure 9.11. Lateral shearing interferometer using two identical gratings (P. Hariharan, W. H. Steel, and J. C. Wyant, *Opt. Commun.* **11**, 317–320, 1974).

9.6.2 Radial Shearing Interferometers

In a radial shearing interferometer, two concentric images of the test wavefront, of different sizes, are superposed as shown in Figure 9.12. The ratio of their diameters is known as the shear ratio (μ).

Typical interferograms obtained for the primary aberrations with a radial shearing interferometer are presented in Figure 9.13, along with the corresponding Twyman–Green interferograms. As can be seen, if the shear ratio is small ($\mu = 0.3$), the radial shearing interferogram is very similar to that obtained with a Twyman–Green or Fizeau interferometer and can be interpreted in the same

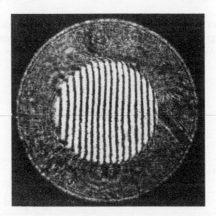

Figure 9.12. Typical interferogram obtained with a radial shearing interferometer (P. Hariharan and D. Sen, *J. Sci. Instrum.* **38**, 428–432, 1961).

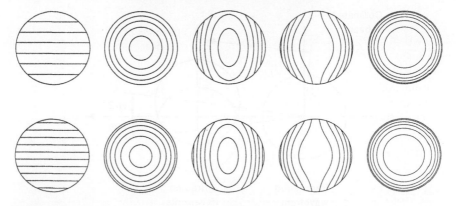

Figure 9.13. Interferograms obtained with a radial shearing interferometer ($\mu = 0.3$) for (left to right) tilt, defocusing, astigmatism, coma, and spherical aberration. The corresponding interferograms obtained with a Twyman–Green interferometer are shown in the lower row.

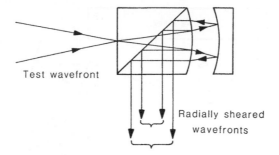

Figure 9.14. Simple radial shearing interferometer (Zhou Wanzhi, *Opt. Commun.* **53**, 74–76, 1985).

manner. The exact values of the aberration coefficients can also be obtained by a procedure similar to that used with Twyman–Green and Fizeau interferograms (see Appendix L.2).

A simple radial shearing interferometer is shown in Figure 9.14. Interference takes place between the wavefronts reflected from the test concave surface and the convex surface of the beam-splitting prism. The radius of the convex surface is chosen to give the desired shear ratio.

9.7 GRAZING-INCIDENCE INTERFEROMETRY

With conventional interferometers, it is not possible to obtain interference fringes with rough surfaces that do not give a specular reflection with visible light at normal incidence.

One way to solve this problem is to use a longer wavelength. Infrared interferometry with a CO_2 laser at a wavelength of 10.6 μm has been used to test ground aspherical surfaces before polishing.

A simpler alternative with nominally flat surfaces is to use visible light incident obliquely on the surface at an angle at which it is specularly reflected. The contour interval is then $\lambda/2\cos\theta$, where θ is the angle of incidence, and interference fringes can be obtained with fine-ground glass and metal surfaces.

The low reflectivity of the test surface can be compensated for by means of a system, such as that shown in Figure 9.15, using a pair of blazed reflection gratings to divide and recombine the beams.

9.8 SUMMARY

- Fizeau and Twyman–Green interferograms are contour maps of the wavefront errors.

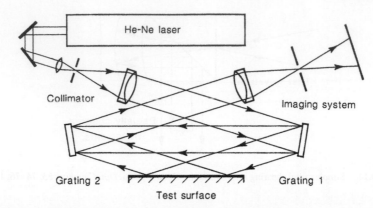

Figure 9.15. Grazing-incidence interferometer using two reflection gratings (P. Hariharan, *Opt. Eng.* **14**, 257–258, 1975).

- Lateral shearing interferograms give the derivative of the wavefront errors in the direction of shear.
- Radial shearing interferograms are very similar to Fizeau and Twyman–Green interferograms.
- Ground surfaces can be tested at grazing incidence, or by using a longer wavelength (a CO_2 laser).

9.9 PROBLEMS

Problem 9.1. *Three nominally flat plates are tested in pairs, in a light box similar to that shown in Figure 9.1, with a mercury vapor lamp ($\lambda = 546$ nm) as the source. The following values are obtained for the deviations of the diametrical fringe from straightness, expressed as fractions of the fringe spacing (a plus sign, within brackets, signifies a contact at the center; a minus sign, a contact at the edge):*

$$A + B = 0.4(+),$$
$$B + C = 0.0,$$
$$C + A = 0.2(-).$$

What are the deviations from flatness of the individual surfaces?

The fringe patterns observed in each case correspond to the algebraic sum of the deviations from flatness of the two plates. Accordingly, we only have to solve the set of equations to obtain the individual errors.

$$A = +0.1 \text{ fringe } (0.05\lambda \text{ convex}),$$

$$B = +0.3 \text{ fringe } (0.15\lambda \text{ convex}),$$

$$C = -0.3 \text{ fringe } (0.15\lambda \text{ concave}). \tag{9.4}$$

Problem 9.2. *The two surfaces of a circular glass plate (n = 1.53) with a diameter of 100 mm are polished flat and nominally parallel. The plate is examined in a Fizeau interferometer from which the reference flat surface has been removed. With an He–Ne laser as the source (λ = 633 nm), straight, parallel interference fringes with a spacing of 12.5 mm are seen. What is the angle between the faces of the plate? If this plate is introduced in one beam of a Twyman–Green interferometer adjusted to produce a uniform field, what would be the separation of the fringes?*

The change in the thickness of the plate corresponding to successive fringes in the Fizeau interferometer is

$$\Delta d_F = \lambda/2n$$

$$= 0.633 \times 10^{-6}/2 \times 1.53$$

$$= 0.207 \ \mu\text{m}. \tag{9.5}$$

Accordingly, the angle between the faces of the plate is

$$\epsilon = 0.207 \times 10^{-6}/12.5 \times 10^{-3}$$

$$= 16.56 \times 10^{-6} \text{ radian}$$

$$= 3.42 \text{ arc sec.} \tag{9.6}$$

With the Twyman–Green interferometer, the change in thickness between successive fringes is

$$\Delta d_T = \lambda/2(n - 1)$$

$$= 0.597 \ \mu\text{m}. \tag{9.7}$$

Accordingly, the spacing of the fringes in the Twyman–Green interferometer would be

$$\Delta x = \Delta d_T/\epsilon$$

$$= 0.597 \times 10^{-6}/16.56 \times 10^{-6}$$

$$= 36.1 \text{ mm}. \tag{9.8}$$

Problem 9.3. *What should be the diameter of the pinhole in a point-diffraction interferometer intended to test a telescope objective with an aperture of 150 mm and a focal length of 2250 mm?*

If we assume that the light from the star has a mean wavelength of 550 nm, the diameter of the diffraction-limited image of the star formed by the telescope objective would be, from Eq. 6.7,

$$d = 2.44 \times 550 \times 10^{-9} \times 2.25/0.15$$

$$= 20.2 \ \mu\text{m}. \tag{9.9}$$

To produce a spherical diffracted wavefront, free from aberrations, the pinhole should be significantly smaller than 20 μm. A 10-μm pinhole should give satisfactory results.

Problem 9.4. *Derive an expression for the optical path difference in the interferogram produced by a lateral shearing interferometer, for a test wavefront having the primary aberrations specified by Eq. 9.2, when the shear is very small.*

For a small shear Δx along the x direction, the optical path difference in the interferogram is proportional to the derivative of the wavefront errors along the x direction. We then have

$$\Delta W_x = \Delta x (\partial W/\partial x)$$

$$= \Delta x \left[A \left(4x^3 + 4xy^2 \right) + 2Bxy + 2Cx + 2Dx + F \right]. \tag{9.10}$$

For a small shear Δy along the y direction, the optical path difference in the interferogram is

$$\Delta W_y = \Delta y (\partial W/\partial y)$$

$$= \Delta y \left[A \left(4y^3 + 4x^2 y \right) + 2B \left(x^2 + 3y^2 \right) + 6Cy + 2Dy + E \right]. \tag{9.11}$$

Note that there is a significant difference between the two shearing interferograms.

Problem 9.5. *The beams in a grazing-incidence interferometer used to test nominally flat, fine-ground surfaces make angles with the test surface of $\pm 10.0°$. If the light source is an He–Ne laser ($\lambda = 633$ nm), what is the contour interval?*

The angle of incidence on the test surface is

$$\theta = 90.0 - 10.0 = 80.0°. \tag{9.12}$$

Accordingly, the contour interval is

$$\Delta z = \lambda/2 \cos \theta$$
$$= 633 \times 10^{-9}/2 \times 0.1736$$
$$= 1.82 \; \mu\text{m.} \tag{9.13}$$

FURTHER READING

For more information, see

D. Malacara, *Optical Shop Testing*, John Wiley, New York (1992).

10

Digital Techniques

The interference pattern obtained with a Fizeau or Twyman–Green interferometer is a contour map of the errors of the wavefront. However, the accuracy of measurements on two-beam fringes is typically only around $\lambda/10$; in addition, calculation of the aberration coefficients from measurements on such an interference pattern is tedious and time-consuming. More accurate measurements can be made and calculations speeded up by using digital image processing techniques.

Some techniques that we will discuss are

- Digital fringe analysis
- Digital phase measurements
- Tests on aspheric surfaces

10.1 DIGITAL FRINGE ANALYSIS

A typical digital system for fringe analysis uses a television camera in conjunction with a video frame memory to measure and store the intensity distribution in the interference pattern; this information is then processed in a computer to locate the intensity maxima and minima. Since the interference fringes do not contain explicit information on the sign of the errors, a tilt is introduced between the interfering wavefronts, so that a linear phase gradient is added to the actual phase differences being measured. Nominally straight fringes are obtained, whose shape is modified by the wavefront errors. Fourier analysis of the fringes then yields unambiguous values of the wavefront errors.

10.2 DIGITAL PHASE MEASUREMENTS

Direct measurements of the optical path difference between the two interfering wavefronts by electronic techniques offer many advantages. Measurements can be made quickly, and with very high accuracy ($\lambda/100$, or better), at a uniformly spaced array of points covering the interference pattern, and the sign of the error can be determined without any ambiguity.

Two methods are commonly used for digital phase measurements. In one method (phase shifting), the optical path difference between the interfering beams is varied linearly with time, as shown in Figure 10.1(a), and the output current from a photodetector located at a point on the interference pattern is integrated over a number of equal segments covering one period of the sinusoidal output signal. In another method (phase stepping), the optical path difference between the interfering wavefronts is changed in equal steps, as shown in Figure 10.1(b), and the corresponding values of the intensity are measured. From the standpoint of theory, the two methods are equivalent. Four measurements at each point can be used conveniently to calculate the original phase difference between the wavefronts at this point (see Appendix M). A charge-coupled detector (CCD) array can be used to make measurements simultaneously at a very large number of points (typically, 512×512) covering the interference pattern. Since the phase calculation algorithm only yields phase data to modulo 2π, subsequent processing is usually necessary to link adjacent points and remove discontinuities.

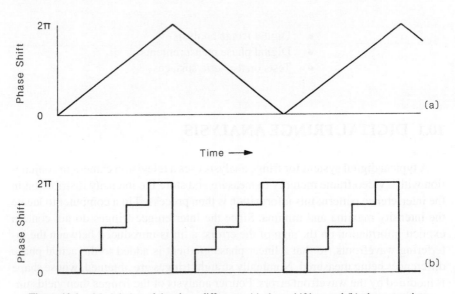

Figure 10.1. Modulation of the phase difference: (a) phase shifting, and (b) phase stepping.

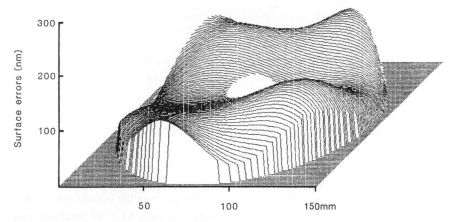

Figure 10.2. Three-dimensional plot of the errors of a spherical concave surface obtained with a digital phase-shifting interferometer.

The simplest way to generate the phase shifts (or phase steps) is to mount one of the mirrors of the interferometer on a piezoelectric transducer (PZT) to which appropriate voltages are applied. Another way is to use a diode laser whose output wavelength can be changed by varying the injection current. If the optical path difference between the two beams in the interferometer is initially set at a known value, say p, the additional phase difference $\Delta\phi$ introduced between the beams by a wavelength change $\Delta\lambda$ is given by the relation

$$\Delta\phi \approx 2\pi p\left(\frac{\Delta\lambda}{\lambda^2}\right). \tag{10.1}$$

Figure 10.2 shows a three-dimensional plot of the errors of a concave surface produced by an interferometer with a digital phase measurement system.

Because of their speed and accuracy, digital interferometers are used extensively in the production of high-precision optical components. New tests are also practical with computerized acquisition and processing of the data. One example is subaperture testing, in which a large reference surface is effectively synthesized from observations made with a small reference surface moved into different positions over the test aperture. Another is testing surfaces with significant deviations from a sphere (aspheric surfaces).

10.3 TESTING ASPHERIC SURFACES

In this section, we will discuss the following topics:

- Direct measurements of surface shape
- Long-wavelength tests
- Tests with shearing interferometers
- Tests with computer-generated holograms

10.3.1 Direct Measurements of Surface Shape

With a digital interferometer, the simplest method of testing surfaces with small deviations from a sphere is to generate a table giving the theoretical deviations of the wavefront, at the measurement points, from the best-fit sphere, and to subtract these values from the corresponding readings. As in all mapping tests with aspheres, care should be taken to image the surface under test on the CCD array to avoid errors.

10.3.2 Long-Wavelength Tests

Direct measurements of surface shape become difficult with surfaces having large deviations from a sphere, when the fringe spacing becomes comparable to that of the detector elements. One method of testing such surfaces is to use a longer wavelength to reduce the number of fringes in the interferogram. Typically, a carbon dioxide laser operating at a wavelength of 10.6 μm can be used, with a pyroelectric vidicon as the detector. An advantage, here, is that the surface can be tested even in the fine-ground state, before it is polished.

10.3.3 Tests with Shearing Interferometers

Surfaces with large deviations from a sphere can also be tested with a shearing interferometer (see Section 9.6). In a lateral shearing interferometer, two images of the test wavefront are superposed with a small mutual lateral displacement. For a small shear (see Appendix L.1), the optical path difference at any point in the interference pattern corresponds to the derivative of the wavefront errors (i.e., the errors in the slope of the test surface), and the sensitivity can be varied by adjusting the amount of shear. Evaluation of the wavefront aberrations is easier with a radial shearing interferometer, in which interference takes place between two differently sized images of the test wavefront. In this case, the number of fringes in the interferogram can be reduced by making the difference in the diameters of the two images of the test wavefront very small. The phase data can then be processed readily to obtain the shape of the test wavefront (see Appendix L.2).

10.3.4 Tests with Computer-Generated Holograms

Null tests are preferable for surfaces with very large deviations from a sphere, since the requirement for exact imaging of the surface is then less critical. One method is to use a suitably designed null lens that converts the wavefront leaving the surface under test into an approximately spherical wavefront. A more flexible alternative, that is now used widely, is a computer-generated hologram (CGH).

Figure 10.3 is a schematic of a setup using a CGH in conjunction with a Twyman–Green interferometer to test an aspheric mirror. The CGH resembles the interference pattern formed by the wavefront from an aspheric surface with the specified profile and a tilted plane wavefront, and is positioned so that the mirror under test is imaged onto it. The deviation of the surface under test from its specified shape is then given by the moire pattern formed by the actual interference fringes and the CGH, which is isolated by means of a small aperture placed in the focal plane of the imaging lens.

Figure 10.4(a) shows the distorted fringe patterns obtained with an aspheric surface when tested normally, while Figure 10.4(b) shows the corrected fringe pattern obtained with the CGH specified for it.

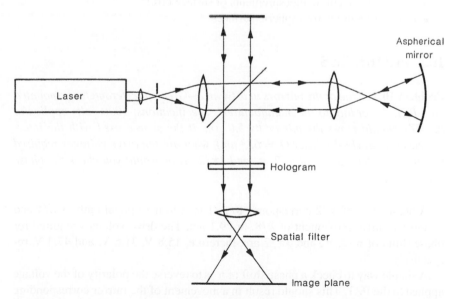

Figure 10.3. Modified Twyman–Green interferometer using a computer-generated hologram (CGH) to test an aspherical mirror (J. C. Wyant and V. P. Bennett, *Appl. Opt.* **11**, 2833–2839, 1972).

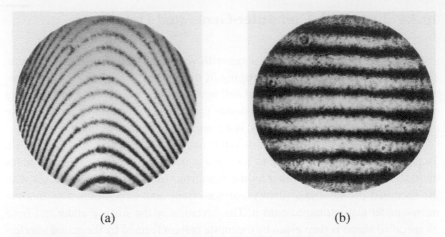

(a)   (b)

Figure 10.4. Interferograms of an aspheric surface (a) without, and (b) with, a compensating computer-generated hologram (CGH) (J. C. Wyant and V. P. Bennett, *Appl. Opt.* **11**, 2833–2839, 1972).

10.4 SUMMARY

Digital techniques permit

- speedy and accurate measurements of surface errors
- new test methods for aspheric surfaces

10.5 PROBLEMS

Problem 10.1. *One of the mirrors in a Twyman–Green interferometer is mounted on a PZT. According to the manufacturer's specifications, 1000 volts applied to the PZT should move the mirror by 5.0 μm. If the source used with the interferometer is an He–Ne laser ($\lambda = 633$ nm), what are the drive voltages required to introduce phase shifts of $\pi/2$, π, and $3\pi/2$? How would you check the phase shifts?*

A phase shift of $\pi/2$ corresponds to a change in the optical path of $\lambda/4$ and requires a mirror movement of $\lambda/8$, or 79.1 nm. The drive voltages required for phase shifts of $\pi/2$, π, and $3\pi/2$ are, therefore, 15.8 V, 31.6 V, and 47.4 V, respectively.

A simple way to check a phase shift of π is to reverse the polarity of the voltage applied to the PZT. This should result in a movement of the mirror corresponding to a phase shift of 2π, which should produce no apparent movement of the fringes in the interferometer.

Problem 10.2. *The light source used in a Fizeau interferometer is a diode laser* ($\lambda = 790$ nm). *A change in the injection current of 1.0 mA shifts the output wavelength by* 8.53×10^{-3} nm. *If the reference and test surfaces in the interferometer are separated by an air gap of 25 mm, what would be the changes in the injection current required to introduce phase shifts of* $\pi/2$, π, *and* $3\pi/2$?

From Eq. 10.1, the wavelength shift required to introduce a phase shift of π is

$$\Delta\lambda = \lambda^2/2p, \tag{10.2}$$

where p is the optical path difference between the two interfering beams. In this case, $p = 2d$, where d is the air gap between the reference and test surfaces in the interferometer, and we have

$$\Delta\lambda = \left(790 \times 10^{-9}\right)^2/50 \times 10^{-3} \text{ m}$$

$$= 12.5 \times 10^{-3} \text{ nm}. \tag{10.3}$$

Accordingly, the changes in the injection current required to produce phase shifts of $\pi/2$, π, and $3\pi/2$ are 0.73 mA, 1.46 mA, and 2.19 mA, respectively.

Problem 10.3. *The deviations from flatness of the faces of a fused silica disk* (*diameter* $D = 150$ mm, *thickness* $d = 25$ mm) *are measured with a digital Fizeau interferometer using an He–Ne laser as the source. If the coefficient of thermal expansion of fused silica is* $\alpha = 0.5 \times 10^{-6}/°C$, *what is the maximum permissible difference in temperature between the two faces of the disk for the systematic error due to this cause not to exceed* $\lambda/100$?

If there is a difference in temperature ΔT between the two nominally flat faces of the disk, they will take the form of two concentric spheres, as shown in Figure 10.5. The radii of curvature of the two surfaces can be written as R and $R+d$, where, to a first approximation,

$$[D(1 + \alpha\Delta T)]/(R+d) = D/R. \tag{10.4}$$

Accordingly,

$$R = d/\alpha\Delta T. \tag{10.5}$$

The maximum deviation of the surface from its true shape is at its center and is given by the relation

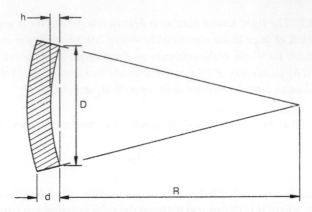

Figure 10.5. Deformation of a disk due to a temperature difference between its two faces.

$$h = D^2/8R$$

$$= D^2\alpha\,\Delta T/8d$$

$$= 0.15^2 \times 0.5 \times 10^{-6}\Delta T/8 \times 0.025$$

$$= 56.3 \times 10^{-9}\Delta T \text{ m.} \tag{10.6}$$

For this deviation not to exceed $\lambda/100$ (6.33×10^{-9} m), the maximum permissible value of the temperature difference ΔT is

$$\Delta T = 6.33 \times 10^{-9}/56.3 \times 10^{-9}$$

$$= 0.11\,°C. \tag{10.7}$$

Problem 10.4. *The same digital Fizeau interferometer is used to test a concave mirror in a setup similar to that shown in Figure 9.3. If the distance from the reference flat to the concave mirror is 400 mm, how well should the laser wavelength be stabilized to ensure that errors due to random fluctuations of the laser wavelength do not exceed $\lambda/50$?*

The measured value of the phase difference at any point in the interferogram is, from Eq. 2.11,

$$\phi = (2\pi/\lambda)p, \tag{10.8}$$

where p is the optical path difference between the two beams in the interferometer. Accordingly, the error $\Delta\phi$ in the measured value of the phase due to a change $\Delta\lambda$ in the laser wavelength is given by the relation

$$\Delta\phi = -\left(2\pi p/\lambda^2\right)\Delta\lambda. \tag{10.9}$$

If the error in the measurements is not to exceed $\lambda/50$, $\Delta\phi$ must be less than $2\pi/50$. The maximum value of $\Delta\lambda$ is then given by the relation

$$
\begin{aligned}
(\Delta\lambda/\lambda) &= \lambda/50p \\
&= 0.633 \times 10^{-6}/50 \times 0.8 \\
&= 1.6 \times 10^{-8}.
\end{aligned}
\tag{10.10}
$$

The stability of the laser wavelength must be better than this figure over the period of data acquisition, typically about a second.

FURTHER READING

For more information, see

1. K. Creath, *Phase-Measurement Interferometry Techniques*, in Progress in Optics, Vol. XXVI, Ed. E. Wolf, North-Holland, Amsterdam (1988), pp. 350–393.
2. D. Malacara, *Optical Shop Testing*, John Wiley, New York (1992).
3. D. W. Robinson and G. T. Reid, *Interferogram Analysis: Digital Techniques for Fringe Pattern Measurements*, IOP Publishers, London (1993).
4. D. Malacara, M. Servin, and Z. Malacara, *Interferogram Analysis for Optical Testing*, Marcel Dekker, New York (1998).

11

Macro- and Micro-Interferometry

Many applications of optical interferometry involve measurements of local variations of shape or refractive index. The topics we will discuss in this chapter are

- Interferometry of refractive index fields
- The Mach–Zehnder interferometer
- Interference microscopy
- Multiple-beam interferometry
- Two-beam interference microscopes
- The Nomarski interferometer

11.1 INTERFEROMETRY OF REFRACTIVE INDEX FIELDS

Interferometry is widely used in studies of fluid flow, combustion, heat transfer, plasmas, and diffusion, where local variations in the refractive index can be related to changes in the pressure, the temperature, or the relative concentration of different components. The Mach–Zehnder interferometer is commonly used for such studies.

11.2 THE MACH–ZEHNDER INTERFEROMETER

A typical setup using the Mach–Zehnder interferometer is shown schematically in Figure 11.1. The Mach–Zehnder interferometer has the advantages that

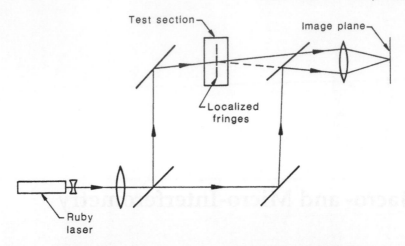

Figure 11.1. Test setup for measurements on refractive index fields with a Mach–Zehnder interferometer.

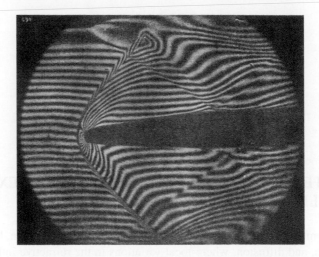

Figure 11.2. Interference pattern of the supersonic flow around an airfoil obtained with a Mach–Zehnder interferometer (R. Chevalerias, Y. Latron, and C. Veret, *J. Opt. Soc. Am.* **47**, 703–706, 1957).

the separation of the two beams can be as large as desired, and the test section is traversed only once. In addition, white-light fringes can be obtained and localized in the same plane as the test section (see Section 3.5). This makes it possible to use a high energy pulsed laser, which may be operating in more than one mode, or a flash lamp, to record interferograms of transient phenomena; it also makes

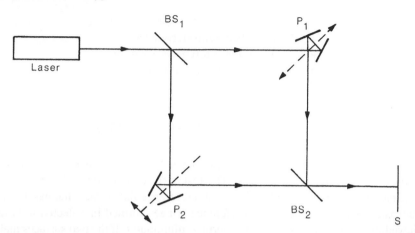

Figure 11.3. Modified Mach–Zehnder interferometer (P. Hariharan, *Appl. Opt.* **8**, 1925–1926, 1969).

it possible, as shown in Figure 11.2, to have the interference fringes and the test section in focus at the same time.

The adjustment of the Mach–Zehnder interferometer to obtain white-light fringes localized in a particular plane usually involves a series of successive approximations and can be quite time-consuming. A systematic procedure is outlined in Appendix G.

For small apertures, it is convenient to use the modified optical arrangement shown in Figure 11.3, in which the two mirrors are replaced by mirror pairs, P_1, P_2, which deviate the beams through a fixed angle. P_1 can be moved along a direction parallel to the plane of symmetry, while P_2 can be moved along a direction perpendicular to this plane. With this arrangement, the optical path difference and the plane of localization of the interference fringes can be controlled independently.

11.3 INTERFERENCE MICROSCOPY

An important application of optical interferometry is in microscopy. Interference microscopy provides a noncontact method for studies of the structure and measurements of the roughness of specular surfaces, when stylus profiling cannot be used because of the risk of damage.

In the next three sections we will discuss the following techniques in interference microscopy:

- Multiple-beam interferometry
- Two-beam interference microscopes
- The Nomarski interferometer

11.4 MULTIPLE-BEAM INTERFEROMETRY

An important application of multiple-beam fringes of equal thickness is in studies of the structure of surfaces. The test surface is coated with a highly reflecting layer of silver and placed against a reference flat surface that has a semitransparent silver coating. The interference fringes formed in reflected light are viewed through a microscope with a vertical illuminator. If the two surfaces make a small angle with each other, the fringes of equal thickness that are formed are effectively profiles of the surface. To obtain the sharpest fringes, it is necessary to ensure that the separation of the surfaces is less than a few micrometres, and the angle between them is very small. The reason for this is that, with a wedged air film, waves formed by successive reflections emerge at progressively increasing angles to the directly reflected wave. As a result, the optical path difference between successive waves is not exactly the same. If this deviation from equality is significant, the intensity distribution in the interference fringes becomes asymmetrical, and their width increases. Typically, for $\lambda = 633$ nm and surfaces with a reflectance of 0.9 separated by 5 μm, the spacing of the fringes should not be less than 1 mm.

These problems can be avoided by using multiple-beam fringes of equal chromatic order (FECO fringes: see Section 5.4). In this case the two surfaces are parallel, so that very sharp fringes can be obtained. As shown in Figure 11.4, FECO fringes can reveal surface irregularities <1 nm high.

FECO fringes yield very high sensitivity with a simple setup. However, if the test surface does not have a high reflectance, it must be coated with a highly reflecting film.

11.5 TWO-BEAM INTERFERENCE MICROSCOPES

Two-beam interference microscopes are available using optical systems similar to the Fizeau, Michelson, and Mach–Zehnder interferometers. Very accurate measurements can be made by phase-shifting (see Appendix M). Measurements can also be made on vibrating parts (MEMS) by using stroboscopic illumination.

The Mirau interferometer, shown in Figure 11.5, permits a very compact setup. In this arrangement, light from an illuminator is incident, through the microscope

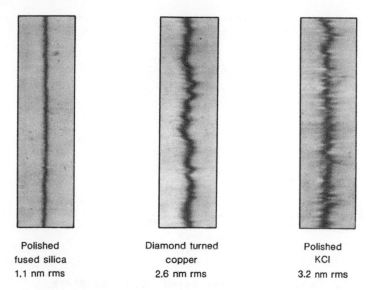

Polished	Diamond turned	Polished
fused silica	copper	KCl
1.1 nm rms	2.6 nm rms	3.2 nm rms

Figure 11.4. FECO fringes showing the residual irregularities of polished surfaces: (left to right) polished fused silica, diamond-turned copper, and polished potassium chloride (courtesy J. M. Bennett, Michelson Laboratory).

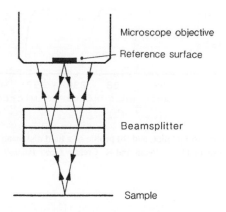

Figure 11.5. The Mirau interferometer.

objective, on a beam splitter. The transmitted beam goes to the test surface, while the reflected beam goes to an aluminized spot on the flat front surface of the microscope objective. The two reflected beams are recombined at the same beam splitter and return through the objective. The interference pattern formed in the

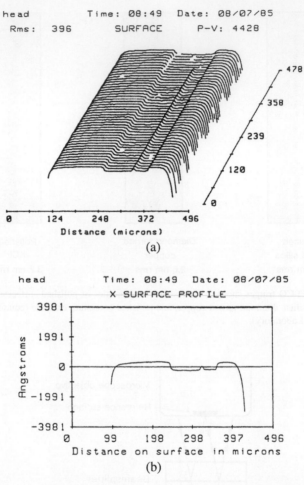

(a)

(b)

Figure 11.6.   (a) Three-dimensional plot, and (b) profile of a hard disk head, obtained with a phase-shifting interference microscope (J. C. Wyant and K. Creath, *Laser Focus/Electro Optics*, 118–132, Nov. 1985).

image plane contours the deviations from flatness of the test surface. As shown in Figure 11.6, very accurate measurements of surface profiles (to better than 1 nm) can be made using digital phase shifting. In the case of measurements on rough surfaces, the data can be processed to plot a histogram of the surface deviations, or to obtain the rms surface roughness and the autocovariance function of the surface deviations.

11.6 THE NOMARSKI INTERFEROMETER

The Nomarski interferometer, shown schematically in Figure 11.7, is a lateral shearing interferometer that uses two Wollaston (polarizing) prisms to split and recombine the beams.

Two methods of observation are possible. With small isolated objects, it is convenient to use a lateral shear larger than the dimensions of the object. Two images of the object are then seen, covered with interference fringes that contour the phase changes due to the object. More commonly, the shear is made much smaller than the dimensions of the object (differential interference contrast microscopy). The interference pattern then shows the phase gradients, and edges are enhanced. As shown in Figure 11.8, this makes it very easy to detect grain structure and local defects, such as scratches.

Since the lengths of the two optical paths are very nearly equal, it is possible to use a white-light source with the Nomarski interferometer. Very small surface irregularities are then revealed by changes in color. In addition, ambiguities arising at steps can be resolved, since corresponding fringes on either side of the step can be identified easily.

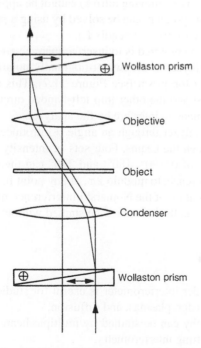

Figure 11.7. The Nomarski interferometer (transmission version).

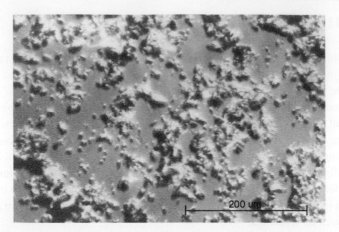

Figure 11.8. Nomarski interference micrograph of a partially polished glass surface, showing remaining grinding pits (J. M. Bennett and L. Mattson, *Introduction to Surface Roughness and Scattering*, Optical Society of America, Washington, DC, 1989).

A problem with the Nomarski interferometer is that conventional phase-shifting techniques (such as a moving mirror) cannot be applied to make quantitative measurements. This problem can be solved by using a phase shifter operating on the Pancharatnam phase (see Appendix E).

The only modification required is to insert a quarter-wave retarder, oriented at 45°, just below the analyzer in the two orthogonally polarized beams emerging from the second Wollaston prism (see Figure 11.7). This retarder converts one beam into right-handed and the other into left-handed circularly polarized light. These two beams are made to interfere by the analyzer.

A rotation of the analyzer through an angle θ introduces an additional phase difference of 2θ between the beams. Four sets of intensity values recorded with additional phase shifts of 0°, 90°, 180°, and 270° can then be used to calculate the original phase difference, to modulo 2π, at any point in the image.

An important application of the Nomarski interference microscope is for studies of transparent living cells that cannot be stained without damaging them.

11.7 SUMMARY

- The Mach–Zehnder interferometer is useful for studies of fluid flow, combustion, heat transfer, plasmas, and diffusion.
- Surface topography can be studied by multiple-beam interferometry or by digital phase-shifting interferometry.
- The Mirau interferometer permits a very compact setup.

- Very small local defects can be seen with the Nomarski interferometer.
- Phase-shifting techniques can be applied to the Nomarski interferometer by using a phase shifter operating on the Pancharatnam phase.
- An important application of the Nomarski interference microscope is for studies of transparent living cells.

11.8 PROBLEMS

Problem 11.1. *A 200-mm-thick test cell in one arm of a Mach–Zehnder interferometer is used in a study of heat transfer by convection. How many interference fringes would you expect to see across the field for a temperature difference of 10°C between the top and bottom of the test cell?*

The refractive index of air at a temperature T °C is

$$n_T = 1 + \left[(n_0 - 1)/(1 + \alpha T)\right](P/101{,}325), \tag{11.1}$$

where

$$n_0 = \text{the refractive index at } 0°\text{C},$$

$$\alpha = 1/273, \text{ the temperature coefficient of expansion},$$

$$P = \text{the pressure in Pascals}.$$

If we take $n_0 = 1.000292$, the change in the refractive index, at constant pressure, for a change in temperature of 10°C is

$$\Delta n = 0.000292 \times 10/273$$

$$= 0.0000107. \tag{11.2}$$

With a test cell having a thickness d, the resulting change in the optical path difference would be

$$\Delta p = d\,\Delta n$$

$$= 2 \times 10^{-1} \times 1.07 \times 10^{-5} \text{ m}$$

$$= 2.14 \; \mu\text{m}. \tag{11.3}$$

At an effective wavelength of 0.55 μm, the number of interference fringes seen across the field would be

$$N = \Delta p / \lambda$$
$$= 2.14/0.55$$
$$= 3.9. \tag{11.4}$$

Problem 11.2. *What would be the smallest step height that could be resolved with a digital phase-shifting interferometer using a 10-bit A-D converter?*

With a 10-bit A-D converter, it should be possible to obtain a phase resolution

$$\Delta\phi \approx 2\pi/2^{10}, \tag{11.5}$$

which, at a wavelength $\lambda = 550$ nm, would correspond to an optical path difference

$$\Delta p = (\lambda/2\pi)\Delta\phi$$
$$= 550/2^{10}$$
$$= 0.54 \text{ nm}, \tag{11.6}$$

or a step height of 0.27 nm.

Problem 11.3. *The lateral resolution that can be obtained in direct measurements with an instrument using a diamond stylus (tip radius r) on a surface with irregularities having an amplitude a is $\Delta x = 2\pi(ar)^{1/2}$. For a surface exhibiting irregularities with an amplitude of 10 nm, what is the lateral resolution that can be obtained with a stylus radius of 0.1 μm? How does this compare with the lateral resolution that can be obtained with a Mirau interferometer using an 8-mm microscope objective (0.4 NA)?*

The lateral resolution that can be obtained with a diamond stylus with a tip radius of 0.1 μm is

$$\Delta x = 2\pi\left(10^{-8} \times 10^{-7}\right)^{1/2}$$
$$= 0.20 \ \mu\text{m}. \tag{11.7}$$

The lateral resolution obtained with the microscope objective (assuming a mean wavelength $\lambda = 550$ nm) is

$$\Delta x = 1.22\lambda/NA$$
$$= 1.22 \times 550 \times 10^{-9}/0.4$$
$$= 1.68 \ \mu\text{m}.$$

While the resolution in depth that can be obtained by interferometry is comparable to that obtained with a stylus profilometer, the lateral resolution is significantly poorer.

FURTHER READING

For more details, see

1. W. Krug, J. Rienitz, and G. Schulz, *Contributions to Interference Microscopy*, Hilger and Watts, London (1964).
2. S. Tolansky, *Multiple-Beam Interference Microscopy of Metals*, Academic Press, London (1970).
3. M. Françon and S. Mallick, *Polarization Interferometers: Applications in Microscopy and Macroscopy*, Wiley-Interscience, London (1971).
4. J. M. Bennett and L. Mattson, *Introduction to Surface Roughness and Scattering*, Optical Society of America, Washington, DC (1989).

12

White-Light Interference Microscopy

A problem with interferometric profilers using monochromatic light is phase ambiguities arising at discontinuities and steps involving a change in the optical path difference greater than a wavelength. This problem can be overcome by using white light.

In this chapter we will discuss

- White-light interferometry
- White-light phase-shifting microscopy
- Spectrally resolved interferometry
- Coherence-probe microscopy

12.1 WHITE-LIGHT INTERFEROMETRY

With white light, the interference term is appreciable only over a very limited range of depths, because of the short coherence length of the illumination. As a result, a three-dimensional image can be extracted by scanning the object in depth and evaluating the degree of coherence (the fringe visibility) between corresponding pixels in the images of the object and reference planes. This technique is known as vertical-scanning white-light interference microscopy, or coherence-probe microscopy.

Figure 12.1 shows the variations in intensity at a given point in the image as the object is scanned in depth along the z-axis. Each such interference pattern can be processed to obtain the envelope of the intensity variations (the fringe-visibility function). The location of the visibility peak along the scanning (z) axis yields the height of the surface at the corresponding point.

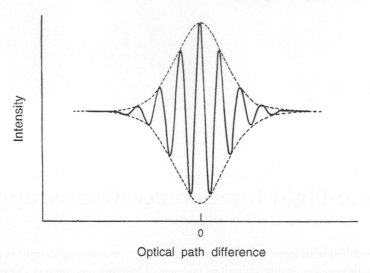

Figure 12.1. Output from an interferometric profiler using white light, as a function of the position of the object along the z-axis.

One way to recover the fringe-visibility function from the sampled intensity data is by digital filtering in the frequency domain. This process involves two Fourier transforms (forward and inverse) along the z direction for each pixel. It is necessary, therefore, for the object to be scanned along the z-axis in steps that correspond to a change in the optical path difference that is less than a fourth of the shortest wavelength. As a result, this procedure requires a large amount of memory and processing time.

12.2 WHITE-LIGHT PHASE-SHIFTING MICROSCOPY

A simpler way to recover the fringe-visibility function is by shifting the phase of the reference wave by three or more known amounts at each position along the z-axis, and recording the corresponding values of the intensity; these intensity values can then be used to evaluate the fringe visibility at that step. However, if the phase shifts are introduced by changing the optical path difference between the beams, the resulting phase shift is inversely proportional to the wavelength.

One way to overcome this problem is to use a five- or seven-step algorithm, which is insensitive to deviations in the values of the phase shifts from their nominal values, to calculate the values of the visibility at each step. High accuracy can then be attained by processing the same intensity values to obtain the fractional phase difference between the beams at the step closest to the visibility maximum.

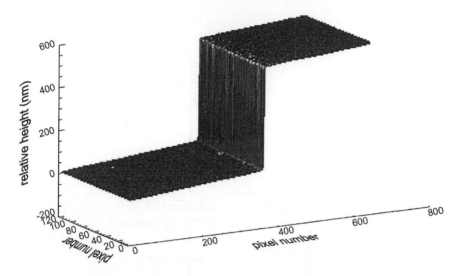

Figure 12.2. Surface profile of a step-height standard obtained with an unfiltered tungsten light source (A. Harasaki, J. Schmit, and J. C. Wyant, *Appl. Opt.* **39**, 2107–2115, 2000).

Figure 12.2 shows a surface profile of a step-height standard obtained by this technique.

Another way to solve this problem is by using an achromatic phase shifter operating on the geometric (Pancharatnam) phase (see Appendix F).

12.3 SPECTRALLY RESOLVED INTERFEROMETRY

Yet another technique that can be used with white light is spectrally resolved interferometry. In this technique, the interferogram is imaged on the slit of a spectroscope, which is used to analyze the light from each point on the slit. The phase difference between the beams, at each point on the object along the line defined by the slit, can then be obtained from the intensity distribution in the resulting channeled spectrum. Higher accuracy can be obtained by phase shifting. The values of the surface height obtained in this manner are free from 2π phase ambiguities. However, each interferogram only yields a profile along a single line.

12.4 COHERENCE-PROBE MICROSCOPY

The optical system of a computer-controlled coherence-probe microscope, which can rapidly and accurately map the shape of surfaces exhibiting steps and discontinuities, is shown schematically in Figure 12.3.

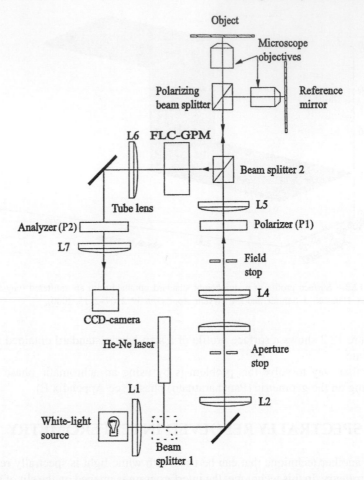

Figure 12.3. Optical system of a coherence-probe microscope using an achromatic phase-shifter operating on the Pancharatnam phase (M. Roy, C. J. R. Sheppard, and P. Hariharan, *Opt. Express* **12**, 2512–2516, 2004).

This instrument used an optical system based on the Linnik interferometer (a modified Michelson interferometer) and scanned the object in height. A switchable achromatic phase shifter operating on the Pancharatnam phase (see Appendix E) was used to evaluate the fringe visibility at each height setting, for an array of points covering the object. The location of the fringe-visibility peak along the scanning axis, for each point on the object, gave the height of the object at the corresponding point. Higher accuracy could then be obtained by processing the intensity data obtained at the step nearest to the visibility peak to yield the fractional phase at this step.

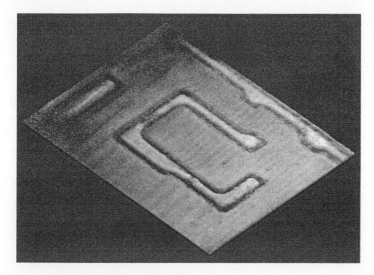

Figure 12.4. Three-dimensional view of an integrated circuit obtained with a coherence-probe microscope. Lateral dimensions of the specimen: 25×43 μm; height about 1 μm (M. Roy, C. J. R. Sheppard, and P. Hariharan, *Opt. Express* **12**, 2512–2516, 2004).

Figure 12.4 shows a three-dimensional view of an integrated circuit obtained with this system.

12.5 SUMMARY

- The use of white light makes it possible to overcome the problem of phase ambiguities at discontinuities and steps.
- The height of the object at any point can be determined by
  - scanning the object in depth and locating the fringe-visibility peak
  - spectrally resolved interferometry.
- High accuracy can be obtained by using phase-shifting techniques.

12.6 PROBLEMS

Problem 12.1. *An achromatic phase shifter operating on the Pancharatnam phase is used with white light in a coherence-probe microscope to shift the phase of the reference beam. Four values of the intensity are recorded at each pixel in the image of the interference pattern, corresponding to additional phase differences of $0°$, $90°$, $180°$, and $270°$. How would you obtain the visibility of the interference pattern from these intensity values?*

With a broadband source, the intensity at any point in the image can be written as

$$I(p, \phi) = I_1 + I_2 + 2(I_1 I_2)^{1/2} g(p) \cos\left[(2\pi/\bar{\lambda})p + \phi_0 + \phi\right], \qquad (12.1)$$

where I_1 and I_2 are the intensities of the two beams acting independently, p is the optical path difference, and $g(p)$ is the corresponding value of the visibility function, $\bar{\lambda}$ is the mean wavelength of the illumination, ϕ_0 is the difference in the phase shifts on reflection at the beam splitter and the mirrors, and ϕ is the additional phase difference introduced by the phase shifter.

Since the additional phase differences introduced are the same for all wavelengths, the corresponding values of the intensity can be written as

$$I(0) = I_1 + I_2 + 2(I_1 I_2)^{1/2} g(p) \cos\left[(2\pi/\bar{\lambda})p + \phi_0\right],$$

$$I(90) = I_1 + I_2 + 2(I_1 I_2)^{1/2} g(p) \sin\left[(2\pi/\bar{\lambda})p + \phi_0\right],$$

$$I(180) = I_1 + I_2 - 2(I_1 I_2)^{1/2} g(p) \cos\left[(2\pi/\bar{\lambda})p + \phi_0\right],$$

$$I(270) = I_1 + I_2 - 2(I_1 I_2)^{1/2} g(p) \sin\left[(2\pi/\bar{\lambda})p + \phi_0\right]. \qquad (12.2)$$

The value of the visibility function $g(p)$ is then given, apart from a normalizing factor, which depends on the relative intensity of the two beams, by the same relation as for monochromatic light,

$$g(p) = \frac{2[(I_0 - I_{180})^2 + (I_{90} - I_{270})^2]^{1/2}}{I_0 + I_{90} + I_{180} + I_{270}}. \qquad (12.3)$$

FURTHER READING

For more information, see

1. P. Hariharan, *Optical Interferometry*, Academic Press, San Diego (2003).
2. P. Hariharan, *The Geometric Phase*, in Progress in Optics, Vol. XLVIII, Ed. E. Wolf, Elsevier, Amsterdam (2005), pp. 149–201.

13

Holographic and Speckle Interferometry

Techniques based on holography and laser speckle can be used to make interferometric measurements on objects with rough surfaces. In this chapter, we will
discuss

- Holographic nondestructive testing
- Holographic strain analysis
- Holographic vibration analysis
- Speckle interferometry
- Electronic speckle-pattern interferometry

13.1 HOLOGRAPHIC INTERFEROMETRY

Holography makes it possible to store and reconstruct a perfect three-
dimensional image of an object (see Appendix N).

When a hologram is replaced in its original position in the recording setup, it
reconstructs the original object wave. If, then, the object undergoes a deformation,
the wavefront from the deformed object will interfere with the wavefront reconstructed by the hologram to produce interference fringes that map the changes in
the shape of the object in real time.

Alternatively, two holograms can be recorded on the same film, one of the
object in its initial condition, and the other of the deformed object. The wavefronts
reconstructed by the two holograms then produce a similar interference pattern.

Holographic interferometry is a very powerful method for mapping changes
in the shape of three-dimensional objects with very high accuracy. Since it is

Figure 13.1. Holographic interferogram of a honeycomb panel, showing sites of poor bonding (D. W. Robinson, *Appl. Opt.* **22**, 2169–2176, 1983).

applicable to objects with rough surfaces, it is used widely for nondestructive testing and strain analysis. It is also very useful for the analysis of vibrations.

13.2 HOLOGRAPHIC NONDESTRUCTIVE TESTING

Holographic interferometry can be used to detect structural weaknesses that produce localized surface deformations when a stress is applied to the object. Typical applications are in detecting cracks and, as shown in Figure 13.1, areas of poor bonding in composite structures.

13.3 HOLOGRAPHIC STRAIN ANALYSIS

To evaluate the strains in an object when it is stressed, it is necessary to measure the actual vector displacements of the surface and differentiate them. As can be seen from Figure 13.2, the phase difference at any point (x, y) in the interferogram is given by the relation

$$\Delta\phi = \vec{L}(x, y) \cdot (\vec{k}_1 - \vec{k}_2)$$
$$= \vec{L}(x, y) \cdot \vec{K}, \tag{13.1}$$

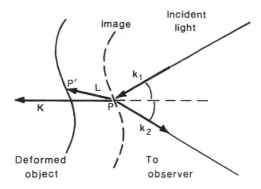

Figure 13.2. Optical path difference produced by a displacement of the object in holographic interferometry.

where $\vec{L}(x, y)$ is the vector displacement of the corresponding point on the surface of the object, $\vec{k}_1$ and $\vec{k}_2$ are vectors of magnitude $k = 2\pi/\lambda$ drawn along the directions of the incident and scattered light, and $\vec{K} = \vec{k}_1 - \vec{k}_2$ is known as the sensitivity vector.

Measurements can be made using an optical system with which four holograms can be recorded in succession, with the object illuminated from two different angles in the vertical plane and two different angles in the horizontal plane. Digital phase-shifting techniques are used to make accurate measurements of the optical path difference at a uniformly spaced network of points. The phase data obtained with the four holograms are then processed to give the vector displacements at these points. These values are used, in conjunction with data on the shape of the object, to calculate the strains.

13.4 HOLOGRAPHIC VIBRATION ANALYSIS

The simplest and most commonly used method for studying vibrating objects is time-average holographic interferometry. In this method, a hologram of the vibrating object is recorded with an exposure time that is much longer than the period of the vibration. The intensity at any point (x, y) in the image is then given by the relation

$$I(x, y) = I_0(x, y) J_0^2 \big[(\vec{k}_1 - \vec{k}_2) \cdot \vec{L}(x, y) \big]$$
$$= I_0(x, y) J_0^2 \big[\vec{K} \cdot \vec{L}(x, y) \big], \tag{13.2}$$

where $I_0(x, y)$ is the intensity with the object stationary, J_0 is the zero-order Bessel function of the first kind, and $\vec{L}(x, y)$ is the amplitude of the vibration

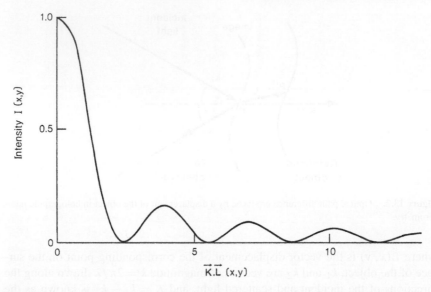

Figure 13.3. Variation of the intensity with the vibration amplitude in the image reconstructed by a time-averaged hologram of a vibrating object.

at that point. The fringes obtained are contours of equal vibration amplitude, with the dark fringes corresponding to the zeros of the function $J_0^2[\vec{K} \cdot \vec{L}(x, y)]$ plotted in Figure 13.3. A series of time-averaged interferograms showing the resonant modes of the soundboard of an acoustic guitar are presented in Figure 13.4.

Another very powerful technique for studies of vibrating objects is stroboscopic holographic interferometry. In this technique, a hologram of the stationary object is recorded, and the real-time interference pattern obtained with the vibrating object is viewed using stroboscopic illumination. The fringes obtained are contours mapping the instantaneous displacement of the object and are similar to those that would be obtained with a static displacement of the object. Phase-shifting techniques can, therefore, be used to make very accurate measurements. Figure 13.5 shows a three-dimensional plot of the instantaneous displacements of a metal plate vibrating at a frequency of 231 Hz.

13.5 SPECKLE INTERFEROMETRY

The image of any object with a rough surface that is illuminated by a laser appears covered with a random granular pattern known as laser speckle (see Appendix O). In speckle interferometry, the speckled image of an object is made

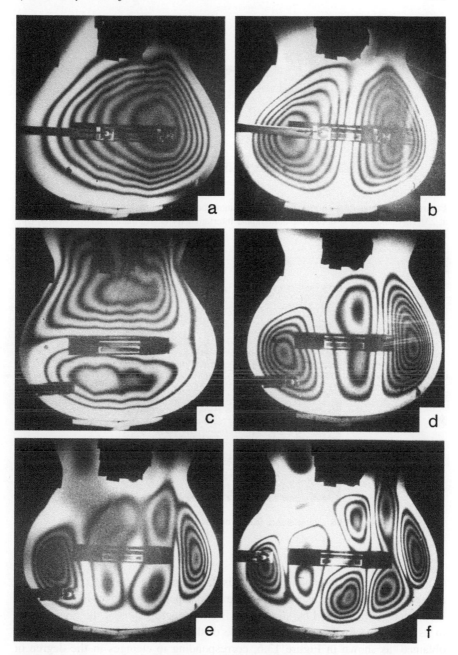

Figure 13.4. Time-average holographic interferograms showing the resonant modes of the sound-board of an acoustic guitar at frequencies of (a) 195, (b) 292, (c) 385, (d) 537, (e) 709, and (f) 905 Hz.

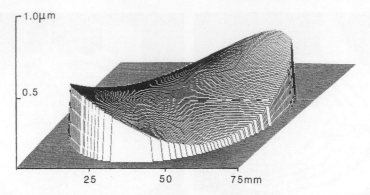

Figure 13.5.   Three-dimensional plot of the instantaneous displacements of a metal plate vibrating at 231 Hz obtained by stroboscopic holographic interferometry (P. Hariharan and B. F. Oreb, *Opt. Commun.* **59**, 83–86, 1986).

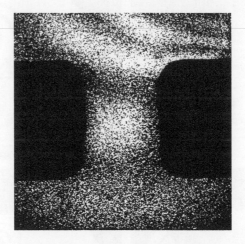

Figure 13.6.   Interference fringes obtained by speckle interferometry.

to interfere with a reference field. Any displacement of the surface then results in changes in the intensity distribution in the speckle pattern. Changes in the shape of the object can be studied by superimposing two photographs of the object taken in its initial and final states. If the shape of the object has changed, fringes are obtained, as shown in Figure 13.6, corresponding to changes in the degree of correlation of the two speckle patterns. These fringes form a contour map of the surface displacements.

13.6 ELECTRONIC SPECKLE-PATTERN INTERFEROMETRY

Electronic speckle-pattern interferometry (ESPI) (also called electronic holographic interferometry) permits very rapid measurements of surface displacements. A typical system used for ESPI is shown in Figure 13.7. The object is imaged on the target of a television camera, along with a coaxial reference beam. The resulting image interferogram has a coarse speckle structure that can be resolved by the television camera.

To measure displacements of the object, an image of the object in its initial state is stored and subtracted from the signal from the television camera. Regions in which the speckle pattern has not changed, corresponding to the condition

$$\vec{K} \cdot \vec{L}(x, y) = 2m\pi, \tag{13.3}$$

where m is an integer, appear dark, while regions where the pattern has changed are covered with bright speckles.

The interference fringes obtained by ESPI are degraded by the coarse speckle pattern covering the image. However, the quality of the fringes can be improved by averaging several identical interference patterns with different speckle backgrounds.

Digital phase-shifting techniques can also be used with ESPI. Each speckle, as seen by the camera, can be regarded as an individual interference pattern, and

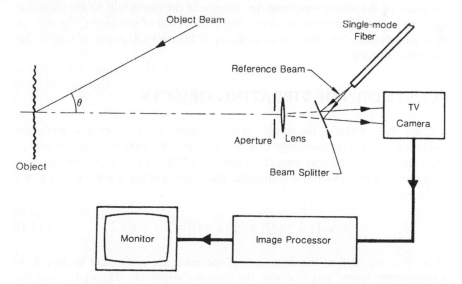

Figure 13.7. System for electronic speckle-pattern interferometry.

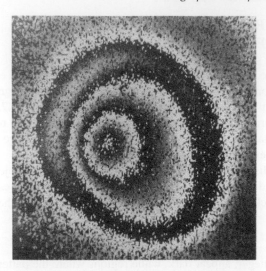

Figure 13.8. Fringes produced by phase-stepping ESPI (D. W. Robinson and D. C. Williams, *Opt. Commun.* **57**, 26–30, 1986).

the phase difference between the beams at this point is measured, by the phase-shifting technique, before and after the object experiences a displacement. The difference of these values is then evaluated. Even though any two speckles may have different initial intensities, corresponding to different values of the amplitude and phase of the object wavefront, the change in the phase will be the same for the same surface displacement. Accordingly, the result of subtracting the second set of phase values from the first is, as shown in Figure 13.8, a contour map of the object deformation.

13.7 STUDIES OF VIBRATING OBJECTS

If the object vibrates, the speckle in the vibrating areas is averaged, while the nodes stand out as regions of high-contrast speckle. If the period of the vibration is small compared to the scan time of the camera, ESPI can be used for quantitative measurements of the vibration amplitude. The contrast of the speckle is then given by the expression

$$C = \left\{1 + 2\beta J_0^2\big[\vec{K} \cdot \vec{L}(x, y)\big]\right\}^{1/2}/(1 + \beta), \tag{13.4}$$

where β is the ratio of the intensities of the reference and object beams, $\vec{K}$ is the sensitivity vector, and $\vec{L}(x, y)$ is the vibration amplitude. The signal from the camera is processed to remove the DC component due to the reference beam, filtered, rectified, and displayed on the monitor. Regions corresponding to the zeros of the Bessel function appear as dark fringes.

Phase-shifting techniques can also be used with ESPI to study vibrating objects. Four sequential television frames are stored, with the phase of the reference beam advanced by 90° between frames. The intensity values for each data point in alternate frames are subtracted from each other. The sum of the squares of these differences gives the magnitude of the function $J_0[\vec{K} \cdot \vec{L}(x, y)]$ at each data point.

ESPI can also be used with stroboscopic illumination, in which case $\cos^2$ fringes are obtained. Stroboscopic illumination even makes it possible to study the vibrations of unstable objects, such as the human ear drum *in vivo*.

13.8 SUMMARY

- Holographic and speckle interferometry permit measurements on objects with rough surfaces.
- Object deformations and vibration amplitudes can be measured very accurately.
- Measurements can be made very rapidly with electronic speckle-pattern interferometry.

13.9 PROBLEMS

Problem 13.1. *A hologram is recorded of a circular diaphragm clamped by its edge over an opening in a pressure vessel and illuminated at 45° with a beam from an He–Ne laser ($\lambda = 633$ nm). The hologram is replaced and viewed in a direction normal to the surface of the diaphragm. When the pressure in the vessel is increased slightly, four concentric circular fringes are seen covering the reconstructed image. What is the deflection of the center of the diaphragm?*

Since the edge of the diaphragm is fixed, and we have four fringes from the center to the edge, the phase difference at the center is

$$\Delta\phi = 8\pi = 25.13 \text{ radians.} \tag{13.5}$$

In addition, it follows, from Figure 13.2 and Eq. 13.1, that the magnitude of the sensitivity vector is

$$K = (2\pi/\lambda)(1 + \cos 45°)$$
$$= 16.94 \times 10^6 \text{ m}^{-1}, \tag{13.6}$$

and it bisects the angle between the directions of illumination and viewing. We also know that the displacement of the center of the diaphragm must be along the normal to its surface. The displacement of the center of the diaphragm is, therefore,

$$L = \Delta\phi / K \cos 22.5°$$
$$= 25.13/15.65 \times 10^6$$
$$= 1.60 \ \mu m. \tag{13.7}$$

Problem 13.2. *If the time-average holograms of the guitar in Figure 13.4 have been recorded with the same setup as that described for Problem 13.1, what would be the vibration amplitudes of the soundboard in the resonant modes at 195 and 292 Hz?*

In these two modes, we have 7 and 6 dark fringes, respectively, from the edge of the soundboard to the point vibrating with the largest amplitude. Since the edge is at rest, the amplitudes of vibration at these points correspond (see Eq. 13.2) to the seventh and sixth zeros of the function $J_0[\vec{K} \cdot \vec{L}(x, y)]$. Accordingly, we have

$$\vec{K} \cdot \vec{L}_{195} = 21.21,$$
$$\vec{K} \cdot \vec{L}_{292} = 18.07. \tag{13.8}$$

The corresponding values of the vibration amplitude are, therefore,

$$\vec{L}_{195} = 21.21/15.65 \times 10^6 = 1.36 \ \mu m,$$
$$\vec{L}_{292} = 18.07/15.65 \times 10^6 = 1.15 \ \mu m. \tag{13.9}$$

Note that, for the mode at 292 Hz, a section of the soundboard between the two peaks is at rest, and the displacements of the two peaks are in opposite senses.

FURTHER READING

For more information, see

1. R. Jones and C. Wykes, *Holographic and Speckle Interferometry*, Cambridge University Press, Cambridge (1989).
2. P. K. Rastogi, Ed., *Holographic Interferometry*, Springer-Verlag, Berlin (1994).
3. P. Hariharan, *Optical Holography*, Cambridge University Press, Cambridge (1996).

14

Interferometric Sensors

Interferometers can be used to measure flow velocities and vibration amplitudes; they can also be used as sensors for several physical quantities and as rotation sensors. Applications being explored include gravitational wave detectors and optical signal processing.

Some of the topics that we will review in this chapter are

- Laser–Doppler interferometry
- Measurements of vibration amplitudes
- Fiber interferometers
- Rotation sensing
- Laser-feedback interferometers
- Gravitational wave detectors
- Optical signal processing

14.1 LASER–DOPPLER INTERFEROMETRY

Laser–Doppler interferometry is now used widely to measure flow velocities. This technique makes use of the fact that light scattered by a moving particle has its frequency shifted by the Doppler effect. This frequency shift can be detected by the beats produced when interference takes place between the scattered light and a reference beam (see Appendix J). Alternatively, the scattered light from two illuminating beams, incident on the moving particle at different angles, can be made to interfere.

A typical optical system using two intersecting laser beams, making angles $\pm\theta$ with the direction of observation, to illuminate the test field is shown in Figure 14.1. Light scattered by a particle passing through the region of overlap of

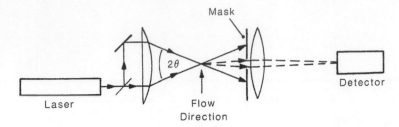

Figure 14.1.   Laser–Doppler interferometer for measurements of flow velocities.

the two beams is focused on a photodetector. If v is the component of the velocity of the particle in the plane of the beams, at right angles to the direction of observation, the frequency of the beat signal is

$$\Delta v = (2v \sin \theta)/\lambda. \tag{14.1}$$

If a small frequency difference is introduced between the two beams, by a pair of acousto-optic modulators operated at slightly different frequencies (see Appendix K), it is possible to distinguish between negative and positive flow directions. Simultaneous measurements of the velocity components along two orthogonal directions can be made with an arrangement using two sets of illuminating beams in orthogonal planes. To avoid interaction between the two pairs of beams, a different laser wavelength is used for each pair of beams.

Laser–Doppler interferometry is also used for noncontact measurements of the velocity of moving surfaces.

14.2 MEASUREMENTS OF VIBRATION AMPLITUDES

Laser interferometry can be used to measure very small vibration amplitudes. Typically, one of the beams in an interferometer is reflected from a mirror attached to the vibrating object. As a result, the frequency of the reflected light is modulated by the Doppler effect. This reflected beam is made to interfere with a reference beam with a fixed frequency offset. The time-varying output from a photodetector then consists (to a first approximation) of a component at the offset frequency (the carrier) and two sidebands (see Appendix P). The vibration amplitude a can then be calculated from the relation

$$2\pi a/\lambda = I_s/I_o, \tag{14.2}$$

where I_o is the power at the offset frequency, and I_s is the power in each of the sidebands.

14.3 FIBER INTERFEROMETERS

Optical fibers are made of a glass core, with a refractive index n_1, surrounded by a cladding with a lower refractive index n_2. A light beam can, therefore, be trapped within the core by total internal reflection. The critical angle i_c at the interface between the core and the cladding is given by the relation

$$\sin i_c = n_2/n_1. \tag{14.3}$$

We can then see, from Figure 14.2, that θ_m, the maximum value of the angle of incidence of a ray on the end of the fiber for it to be trapped within the core, is given by the relation

$$\sin \theta_m = n_1 \cos i_c$$
$$= \left(n_1^2 - n_2^2\right)^{1/2}$$
$$\approx \left[2n_1(n_1 - n_2)\right]^{1/2}. \tag{14.4}$$

The numerical aperture (NA) of the optical fiber is $\sin \theta_m$. Light from a laser focused on the end of the fiber by a microscope objective with an NA equal to or less than this value will be trapped within the fiber and transmitted along the fiber.

It can be shown that only waves at particular angles to the axis are propagated along an optical fiber. These waves correspond to the modes of the fiber. However, if the diameter of the core is less than a few micrometres, the fiber can support only one mode, corresponding to a plane wavefront propagating along the axis of the fiber.

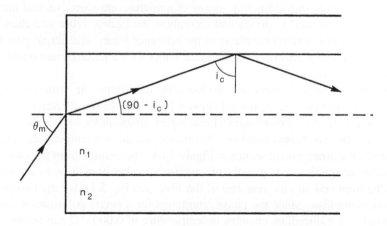

Figure 14.2.   Transmission of light through an optical fiber.

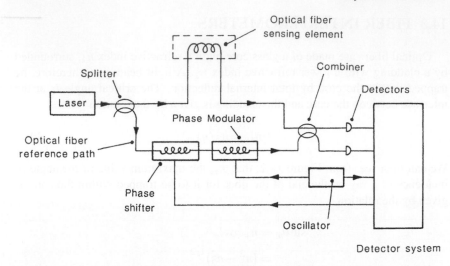

Figure 14.3. Interferometer using a single-mode fiber as a sensing element (T. G. Giallorenzi et al., *IEEE J. Quant. Electron.* **QE18**, 626–665, 1982). ©IEEE, 1982. Reproduced with permission.

Interferometers in which the two beams propagate in single-mode fibers can be used as sensors for a number of physical quantities. Since the optical path length in a fiber changes when it is stretched and is also affected by its temperature, a length of fiber in one arm of the interferometer can be used as a sensing element to measure such changes. Fibers make it possible to have very long, noise-free paths in a small space, so that high sensitivity can be obtained. Figure 14.3 shows a typical optical setup. Light from a diode laser is focused on the cleaved (input) end of a single-mode fiber by means of a microscope objective, and optical fiber couplers are used to divide and recombine the beams. Fiber stretchers are used to shift and modulate the phase of the reference beam. The output goes to a photodetector, and measurements are made either with a heterodyne system or a phase-tracking system.

Birefringent optical fibers are produced by modifying the structure of the cladding, so as to introduce unequal stresses in the core in two directions at right angles to its axis. A section of such a birefringent, single-mode optical fiber, operating as a reflective Fabry–Perot interferometer, is used as a temperature-sensing element in the arrangement shown in Figure 14.4. The outputs from the two photodetectors are processed to give the phase retardation between the waves reflected from the front end and the rear end of the fiber (see Eq. 5.1), for the two polarizations in the fiber. Since the phase retardation for a single polarization can be measured to 1 milliradian, changes in temperature of 0.0005 °C can be detected with a 1-cm-long sensing element. At the same time, since the difference between

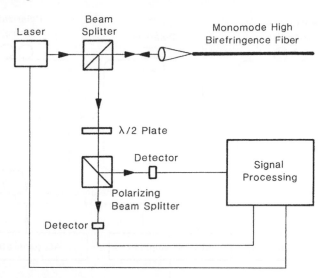

Figure 14.4. Interferometric sensor using a single birefringent monomode fiber (P. A. Leilabady et al., *J. Phys. E* **19**, 143–146, 1986).

the phase retardations for the two polarizations only changes by 2π for a temperature change of $60\,^{\circ}$C, measurements can be made over this entire range.

Fiber interferometers can be used for measurements of magnetic and electric fields by bonding the fiber sensor to a suitable magnetostrictive or piezoelectric element.

Where measurements have to be made of various quantities at a particular location, or a particular quantity at different locations, several fiber-optic sensors can be multiplexed, thereby avoiding duplication of light sources, fiber transmission lines, and photodetectors. Techniques developed for this purpose include frequency-division multiplexing, time-division multiplexing, and coherence multiplexing.

14.4 ROTATION SENSING

Another application of fiber interferometers has been in rotation sensing, where they have the advantages over mechanical gyroscopes of instantaneous response, small size, and relatively low cost.

The arrangement used for this purpose is shown in Figure 14.5 and is the equivalent of a Sagnac interferometer (see Section 3.6). The two waves traverse a closed, multiturn loop, made of a single optical fiber, in opposite directions. If the loop is rotating with an angular velocity Ω about an axis making an angle θ with

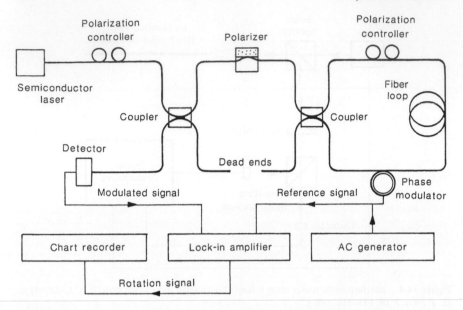

Figure 14.5. Fiber-optic rotation sensor (R. A. Bergh, H. C. Lefevre, and H. J. Shaw, *Opt. Lett.* **6**, 502–504, 1981).

the normal to the plane of the loop, the phase difference introduced between the two waves is

$$\Delta\phi = (4\pi \Omega L r \cos\theta)/\lambda c, \tag{14.5}$$

where L is the length of the fiber, r is the radius of the loop, λ is the wavelength, and c is the speed of light.

14.5 LASER-FEEDBACK INTERFEROMETERS

Laser-feedback interferometers make use of the fact that if, as shown in Figure 14.6, a mirror M_3 is used to reflect the output beam from a laser back into the laser cavity, the laser output varies cyclically with d, the distance of M_3 from M_2, the laser mirror nearest to it.

This effect can be analyzed by considering M_2 and M_3 as a Fabry–Perot interferometer. The reflectance of this interferometer is a maximum when $nd = m\lambda/2$, where n is the refractive index of the intervening medium and m is an integer; it drops to a minimum when $nd = (2m + 1)\lambda/4$. Typically, with a laser mirror having a transmittance of 0.008, the output can be made to vary by a factor of four by using an external mirror with a reflectance of only 0.1.

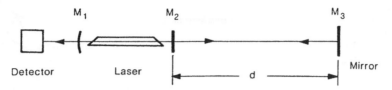

Figure 14.6. Laser-feedback interferometer.

A very compact laser-feedback interferometer can be set up with a single-mode GaAlAs laser and an external mirror mounted on the object whose position is to be monitored.

Since the response of such a system is not linear, its useful range is limited. An increased dynamic range, as well as high sensitivity, can be obtained by mounting the mirror on a PZT and using an active feedback loop to hold the optical path from the laser to the mirror constant.

14.6 GRAVITATIONAL WAVE DETECTORS

The Laser Interferometer Gravitational Observatory (LIGO) project in the USA, and similar projects in other countries, is exploring the use of interferometric techniques for detecting gravitational waves.

With a Michelson interferometer in which the beam splitter and the end reflectors are attached to separate, freely suspended masses, the effect of a gravitational wave would be a change in the difference of the lengths of the optical paths of the two beams. However, to obtain the required sensitivity to strains of 10^{-21} over a bandwidth of a kilohertz, unrealistically long arms would be needed.

Increased sensitivity is obtained by using, as shown in Figure 14.7, two identical Fabry–Perot interferometers ($L = 4$ Km) at right angles to each other, with their mirrors mounted on freely suspended test masses. The frequency of the laser is locked to a transmission peak of one interferometer, while the optical path in the other is continually adjusted so that its transmission peak also coincides with the laser frequency. The corrections applied to the second interferometer are then a measure of the changes in the length of this arm with respect to the first arm.

Since the interferometer is normally adjusted so that observations are made on a dark fringe, to avoid overloading the detector, most of the light is returned toward the source. A further increase in sensitivity can then be obtained by recycling this light back into the interferometer by an extra mirror placed in the input beam.

It is anticipated that the LIGO detectors should be able to detect gravitational waves from a few cosmic events every year.

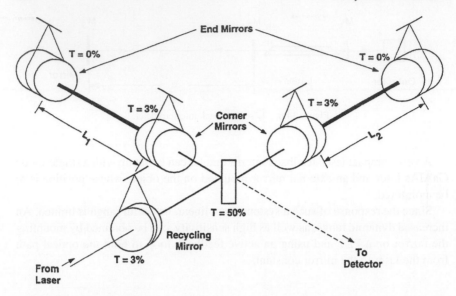

Figure 14.7. Gravitational wave detector using two Fabry–Perot interferometers (R. Weiss, *Rev. Mod. Phys.* **71**, S187–S196, 1999).

14.7 OPTICAL SIGNAL PROCESSING

One area being actively explored is the use of nonlinear optical effects in high-speed interferometric switches and logic gates.

14.7.1 Interferometric Switches

Typically, one of the optical inputs to such a switch is a low-power clock stream (the signal stream) that is split into the two arms of the interferometer and recombined at the output. The other is a high power stream (the control stream) that is used to induce phase changes in one arm. The control stream, which is at a different wavelength from the signal stream, is eliminated by a filter at the output.

For the AND configuration, the interferometer is biased OFF (minimum signal pulse transmission) and the signal pulse is turned ON by the control pulse. In the inverting NOT configuration, the interferometer is biased ON in the absence of the control stream and switched OFF by the control pulse.

14.7.2 Interferometric Logic Gates

Another device that has attracted considerable interest is an optical Boolean XOR gate, which can be used to realize a number of important networking functions. Devices exploiting the nonlinearity of optical fibers have the potential of operating at ultra-high speeds, due to the very short relaxation time (<100 fs) of the nonlinearity, but require long switching lengths and high switching energies. On the other hand, semiconductor optical amplifiers have a nonlinearity that is four orders of magnitude higher than that obtainable with optical fibers, resulting in lower energy requirements and much shorter interaction lengths. Their most important advantage, however, is the possibility of integration to produce compact devices.

Nonlinear effects in semiconductor optical amplifiers have been used to demonstrate XOR logic in various interferometric configurations, opening the way to their exploitation in more complex, all-optical, signal processing circuits.

14.8 SUMMARY

- Flow velocities can be measured by laser–Doppler interferometry.
- Vibration amplitudes can be measured by heterodyne interferometry.
- Fiber interferometers can be used as sensors for pressure, temperature, and electric and magnetic fields, as well as for rotation sensing.
- Laser-feedback interferometers can be used to measure changes in the distance to a target.
- The use of interferometric techniques for detecting gravitational waves is being explored.
- Nonlinear effects are used in high-speed interferometric switches and logic gates.

14.9 PROBLEMS

Problem 14.1. *A laser–Doppler system uses two beams from an Ar^+ laser ($\lambda = 514$ nm), at angles of $\pm 5°$ to the viewing direction, to illuminate the test field. A particle moving across the test field has a velocity component of 1.0 m/sec at right angles to the viewing direction, in the plane of the illuminating beams. What is the frequency of the beat signal?*

From Eq. 14.1, the frequency of the beat signal is

$$\Delta \nu = (2v/\lambda)\sin\theta$$
$$= \left(2/0.514 \times 10^{-6}\right) \times 0.0872$$
$$= 339 \text{ kHz}. \tag{14.6}$$

Problem 14.2. *One of the mirrors in an interferometric setup similar to that described in Section 14.2 is mounted on an ultrasonic transducer. An He–Ne laser ($\lambda = 633$ nm) with an acousto-optic modulator is used as the source. If the transducer is vibrating with an amplitude of 10 nm, what is the power in each of the sidebands relative to that at the carrier frequency?*

From Eq. 14.2, the ratio of the power in each of the sidebands to that at the carrier frequency is

$$I_s/I_o = 2\pi a/\lambda$$
$$= 2\pi \times 10 \times 10^{-9}/633 \times 10^{-9}$$
$$= 0.099. \tag{14.7}$$

Problem 14.3. *A 1.0-m-long fused silica fiber is to be used as the temperature-sensing element in an interferometer. The coefficient of thermal expansion of fused silica is $\alpha = 0.55 \times 10^{-6}/°C$, its refractive index is $n = 1.46$, and the change in refractive index with temperature is $(dn/dT) = 12.8 \times 10^{-6}/°C$. What is the change in the optical path produced by a change in the temperature of the fiber of $1°C$?*

The optical path in a fiber of length L and refractive index n is

$$p = nL. \tag{14.8}$$

The change in the optical path with temperature is, therefore,

$$dp/dt = n(dL/dT) + L(dn/dT)$$
$$= L[n\alpha + (dn/dT)]. \tag{14.9}$$

For a 1-m length of fiber, the change in the optical path for a change in temperature of $1°C$ is

$$\Delta p = 1.46 \times 0.55 \times 10^{-6} + 12.8 \times 10^{-6}$$
$$= 13.60 \ \mu m. \tag{14.10}$$

Problem 14.4. *The smallest phase shift that can be measured in a fiber interferometer used for pressure sensing is 1 μradian. The normalized pressure sensitivity of a typical single-mode fused silica fiber coated with nylon is $\Delta\phi/(\phi\Delta P) = 3.2 \times 10^{-11} \ Pa^{-1}$, where ϕ is the total phase shift produced by the fiber, and*

$\Delta\phi$ *is the change in the phase shift produced by a pressure change* ΔP. *What length of fiber should be used as the sensing element in a marine hydrophone, at a wavelength* $\lambda = 0.633\ \mu m$, *to obtain adequate sensitivity to detect sea-state zero (100 μPa at 1 kHz)?*

To detect sea-state zero, the sensor must have a sensitivity

$$\Delta\phi/\Delta p = 10^{-6}\ \text{radian}/10^{-4}\ \text{Pa}$$
$$= 10^{-2}\ \text{radian/Pa}. \qquad (14.11)$$

This sensitivity can be obtained with a fused silica fiber producing a total phase shift

$$\phi = 10^{-2}/3.2 \times 10^{-11}$$
$$= 3.125 \times 10^8\ \text{radians}. \qquad (14.12)$$

Since the total phase shift produced by a fused silica fiber of length L is

$$\phi = 2\pi n L/\lambda$$
$$= 2\pi \times 1.46L/0.633 \times 10^{-6}$$
$$= 1.449 \times 10^7 L\ \text{radians}, \qquad (14.13)$$

the length of fiber required is

$$L = 3.125 \times 10^8/1.449 \times 10^7$$
$$= 21.6\ \text{m}. \qquad (14.14)$$

Problem 14.5. *A fiber rotation sensor for a navigation application must be capable of detecting a rotation rate equal to 0.1 percent of the earth's rotation rate. If phase measurements can be made with an accuracy of 0.1 μradian at an operating wavelength of 0.85 μm, how many turns of the fiber are required on a 200-mm diameter coil?*

The rotation rate to be detected is

$$\Omega = 10^{-3} \times 2\pi/24 \times 3600$$
$$= 7.27 \times 10^{-8}\ \text{radian/sec}. \qquad (14.15)$$

From Eq. 14.5, the phase shift obtained with a single turn coil for this rotation rate would be

$$\Delta\phi = 4\pi \times 7.27 \times 10^{-8} \times 2\pi \times 0.1^2/0.85 \times 10^{-6} \times 3.00 \times 10^8$$
$$= 2.25 \times 10^{-10} \text{ radian.} \tag{14.16}$$

The number of turns required to obtain a phase shift of 0.1 μradian would be

$$N = 0.1 \times 10^{-6}/2.25 \times 10^{-10}$$
$$= 444 \text{ turns.} \tag{14.17}$$

FURTHER READING

For more information, see

1. R. Durst, A. Melling, and J. H. Whitelaw, *Principles and Practice of Laser–Doppler Anemometry*, Academic Press, London (1981).
2. P. Culshaw, *Optical Fiber Sensing and Signal Processing*, Peregrinus, London (1984).
3. E. Udd, *Fiber Optic Sensors: An Introduction for Engineers and Scientists*, John Wiley, New York (1991).
4. P. Hariharan, *Optical Interferometry*, Academic Press, San Diego (2003).

15

Interference Spectroscopy

Interferometric techniques are now used widely in high-resolution spectroscopy, as well as for wavelength and frequency measurements. Some topics that we will discuss in this chapter are

- Resolving power and etendue
- The Fabry–Perot interferometer
- Interference filters
- Birefringent filters
- Interference wavelength metres
- Laser frequency measurements

15.1 RESOLVING POWER AND ETENDUE

The resolving power of a spectroscope is given by the relation

$$\mathcal{R} = \lambda/\Delta\lambda = \nu/\Delta\nu, \qquad (15.1)$$

where $\Delta\lambda$ or $\Delta\nu$ is the separation of two perfectly monochromatic spectral lines that are just resolved. The resolving power of grating spectroscopes is limited to about 10^6. Higher resolving powers are possible only with interferometers.

Another important characteristic of a spectroscope is its etendue, or throughput. Consider the arrangement shown in Figure 15.1, in which the effective areas A_S and A_D of the source and the detector are images of one another. The amount of radiation accepted by the lens L_S is proportional to $A_S\Omega_S$, where Ω_S is the solid angle subtended by L_S at the source. Similarly, the amount of radiation

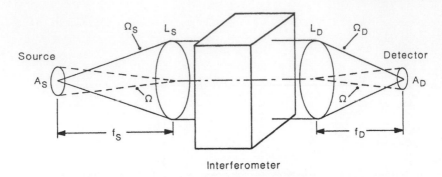

Figure 15.1. Etendue of an interferometer.

reaching the detector is proportional to $A_D\Omega_D$, where Ω_D is the solid angle subtended by L_D at the detector. Since the effective area of the two lenses is the same, the etendue of the instrument is defined by the relation

$$E = A_S\Omega_S = A_D\Omega_D. \tag{15.2}$$

The etendue of a grating spectroscope is limited by the entrance slit, which must be quite narrow for maximum resolution. A much higher etendue can be obtained with an interferometer.

15.2 THE FABRY–PEROT INTERFEROMETER

The Fabry–Perot interferometer (FPI) typically consists of two, slightly wedged, transparent plates with flat surfaces. The inner surfaces of the plates are set parallel to each other and have semitransparent, highly reflecting, multilayer dielectric coatings ($R > 0.95$) (see Section 5.5). The outer surfaces are worked to make a small angle with the inner surfaces, so that reflections from these surfaces can be eliminated. If the spacing of the surfaces is fixed, the interferometer is known as a Fabry–Perot etalon; in this case, a single transparent plate with its surfaces worked flat and parallel, and suitably coated, can also be used.

If the surfaces are separated by a distance d, and the medium between them has a refractive index n, the transmitted intensity, at a wavelength λ, is

$$I_T(\lambda) = T^2 / (1 + R^2 - 2R\cos\phi), \tag{15.3}$$

where $\phi = (4\pi/\lambda)nd\cos\theta$, and θ is the angle of incidence within the interferometer (see Section 5.1). For a given angle of incidence, the difference in the wavelengths (or the frequencies) corresponding to successive peaks in the transmitted intensity (see Figure 5.2) is known as the free spectral range (FSR). From Eq. 5.10,

$$\text{FSR}_\lambda = \lambda^2/2nd,$$

$$\text{FSR}_v = c/2nd. \tag{15.4}$$

The free spectral range corresponds to the range of wavelengths, or frequencies, that can be handled without successive orders overlapping.

The range of wavelengths, or frequencies, corresponding to the width of the peaks (Full Width at Half Maximum, or FWHM) is obtained by dividing the free spectral range by the finesse (see Eq. 5.6) and is given by the relations

$$\Delta\lambda_W = \left(\lambda^2/2nd\right)(1 - R)/\pi R^{1/2},$$

$$\Delta v_W = (c/2nd)(1 - R)/\pi R^{1/2}. \tag{15.5}$$

Accordingly, from Eqs. 15.1 and 15.5, the resolving power of the FPI is

$$\mathcal{R} = (2nd/\lambda)\left[\pi R^{1/2}/(1 - R)\right]. \tag{15.6}$$

If the spacing of the surfaces is fixed (an FP etalon), each wavelength produces a system of rings centered on the normal to the surfaces, as shown in Figure 5.4. With a multiwavelength source, the fringe systems for different spectral lines can be separated by imaging them on the slit of a spectrograph. Each line in the spectrum then contains a narrow strip of the fringe system produced by that line.

15.2.1 The Scanning Fabry–Perot Interferometer

To obtain the full etendue of an FPI, it is necessary to make use of all the rays having the same angle of incidence. This result can be achieved by using the FPI as a scanning spectrometer. In this mode of operation, a small aperture is placed in the focal plane of a lens behind the FPI, and the transmitted intensity is recorded as the spacing of the plates is varied by means of a piezoelectric spacer.

15.2.2 The Confocal Fabry–Perot Interferometer

The etendue of a scanning FPI with plane mirrors is limited by the size of the input and output apertures that can be used without a significant loss in resolution. This limitation is overcome in the confocal Fabry–Perot interferometer which, as shown in Figure 15.2, uses two spherical mirrors whose separation is equal

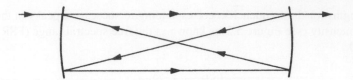

Figure 15.2. Confocal Fabry–Perot interferometer.

to their radius of curvature r, so that their foci coincide. Any incident ray, after traversing the interferometer four times, then emerges along its original path. With this configuration, the optical path difference between successive rays does not depend on the angle of incidence, and a uniform field is obtained. As a result, an extended source can be used, permitting a much higher throughput. A typical application of a scanning confocal FPI is to examine the pattern of longitudinal modes in the output of a cw laser (see Figure 6.2).

15.2.3 The Multiple-Pass Fabry–Perot Interferometer

The contrast factor of an FPI is defined as the ratio of the intensities of the maxima and the minima and is given by the relation

$$C = \left[(1 + R)/(1 - R)\right]^2. \tag{15.7}$$

With typical coatings ($R \approx 0.95$), the background due to a strong spectral line may mask a weak satellite. A higher contrast factor, close to the square of that given by Eq. 15.7, can be obtained by passing the light twice through the same FPI. Contrast factors greater than 10^{10} have been obtained, for studies of Brillouin scattering, with scanning FPIs using up to five passes.

15.3 INTERFERENCE FILTERS

An interference filter can be thought of as a Fabry–Perot interferometer in which the two highly reflecting layers are separated by a thin (1–2 μm thick) spacer layer of a transparent material. Such filters are produced by deposition of the layers, in a vacuum, on a glass substrate. The wavelength for peak transmittance is determined by the thickness of the spacer layer, while the transmission bandwidth depends on the finesse. Two or more identical filters are usually deposited on top of each other to obtain a sharper pass band and lower background transmittance. Unwanted sidebands can be eliminated by a colored glass filter.

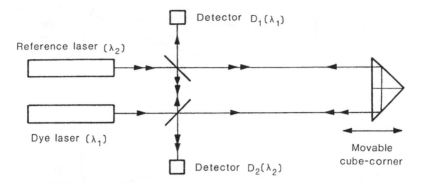

Figure 15.3. Dynamic wavelength metre (V. F. Kowalski, R. T. Hawkins, and A. L. Schawlow, *J. Opt. Soc. Am.* **66**, 965–966, 1976).

15.4 BIREFRINGENT FILTERS

A plate of a birefringent material, such as quartz or calcite (refractive indices n_o and n_e, and thickness d), cut with its faces parallel to the optic axis and set between parallel polarizers, with its optic axis at 45° to the plane of polarization, has a transmission function

$$a(v) = (1/2)\cos^2 \pi v \tau, \tag{15.8}$$

where $\tau = (n_o - n_e)d/c$ is the delay introduced between the two polarizations by the birefringent plate. The transmission peaks corresponding to Eq. 15.8 can be made narrower by using a number of such filters in series, each one with a delay twice that of the preceding one. Birefringent filters were first developed for studies of the surface of the sun; they are now also used as wavelength-selection elements in tunable dye lasers.

15.5 INTERFERENCE WAVELENGTH METRES

Interference wavelength metres are widely used with tunable dye lasers. Dynamic wavelength metres use a two-beam interferometer in which the number of fringes crossing the field is counted as the optical path is changed by a known amount. As shown in Figure 15.3, two beams, one from the dye laser whose wavelength λ_1 is to be determined, and the other from a reference laser whose wavelength λ_2 is known, traverse the same two paths, and the fringe systems formed by the two wavelengths are imaged on separate detectors. If the interferometer is operated in a vacuum, the wavelength of the dye laser can be determined directly

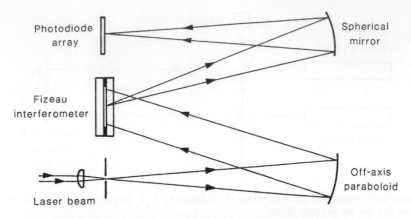

Figure 15.4.   Wavelength metre using a Fizeau interferometer.

by counting fringes simultaneously at both wavelengths, as one end reflector is moved.

A simple static wavelength metre, using a single wedged air film (a Fizeau interferometer), is shown in Figure 15.4. Light from the dye laser forms fringes of equal thickness, whose intensity distribution is recorded directly by a linear photodiode array. The spacing of the fringes can be used to evaluate the integer part of the interference order, while the fractional part can be calculated from the positions of the minima and the maxima with respect to a reference point on the wedge.

15.6 LASER FREQUENCY MEASUREMENTS

As mentioned in Section 8.1, measurements of length are carried out by interferometry, using the vacuum wavelengths of lasers whose frequencies have been compared with the ^{133}Cs standard. The standard procedure for such comparisons used, for many years, a frequency chain of stabilized lasers and nonlinear mixers to bridge the gap between the optical and microwave frequencies.

A completely new approach uses a frequency comb generated by a mode-locked laser. Several modes contribute to the train of pulses produced by such a laser, and the output can be regarded as consisting of a carrier frequency f_c modulated by an envelope function $A(t)$. Fourier transformation of $A(t)$ shows that the resulting spectrum consists of a comb of laser modes, separated by the pulse repetition frequency f_r (the reciprocal of the cavity round-trip time) and centered at the carrier frequency f_c. These modes are very evenly spaced, to a few parts in 10^{17}. The spectral width of the comb can be broadened to extend

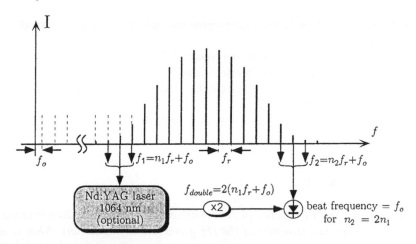

Figure 15.5. Optical frequency comb. The beat frequency between a frequency-doubled "red" component and a "blue" component yields the frequency offset f_o (R. Holzwarth, Th. Udem, T. W. Hänsch, J. C. Knight, W. J. Wadsworth, and P. St. J. Russell, *Phys. Rev. Lett.* **85**, 2264–2267, 2000).

over more than one octave, while maintaining the mode spacing, by sending the pulses through a nonlinear material.

Since f_c is not necessarily an integral multiple of f_r, the modes are shifted by an offset f_o, so that the frequency of the nth mode is

$$f_n = nf_r + f_o, \tag{15.9}$$

where n is a large integer.

The repetition frequency f_r is readily measurable. In addition, with a frequency comb extending over more than one octave, it is possible, as shown in Figure 15.5, to obtain the frequency offset f_o from measurements of the beat frequency between a frequency-doubled "red" component and a properly chosen "blue" component. It is then possible to relate all the optical frequencies f_n, within the comb, to the radio frequencies f_r and f_o, which may be locked to an atomic clock. This allows a direct comparison of an optical frequency with the frequency of an atomic clock, with a precision of 5×10^{-16}.

15.7 SUMMARY

- Interference spectroscopes combine high resolution and high throughput.
- Maximum throughput is obtained with the confocal Fabry–Perot interferometer (FPI).

- A confocal FPI can be used to examine the pattern of longitudinal modes in the output of a cw laser.
- Weak satellites close to strong spectral lines can be studied with the multiple-pass Fabry–Perot interferometer.
- Interference wavelength metres can be used for accurate measurements of dyc lascr wavelengths.
- Extremely accurate measurements of laser frequencies can be made with a frequency comb.

15.8 PROBLEMS

Problem 15.1. *An FPI with two plates separated by an air gap of 20 mm is used to study the hyperfine structure of the Hg green line ($\lambda = 546$ nm). What is its free spectral range? If the reflectance of the surfaces is $R = 0.90$, what is (a) the finesse, (b) the width (FWHM) of the peaks, and (c) the resolving power of the FPI?*

We use Eq. 15.4 to calculate the free spectral range:

$$\text{FSR}_\lambda = 0.007\,45 \text{ nm},$$

$$\text{FSR}_\nu = 7.495 \text{ GHz}, \tag{15.10}$$

while the finesse obtained from Eq. 5.6 is

$$F = 29.8. \tag{15.11}$$

The FWHM of the peaks is obtained by dividing the free spectral range by the finesse or, directly, from Eq. 15.5, and is given by the relations:

$$\Delta\lambda_W = 0.000250 \text{ nm},$$

$$\Delta\nu_W = 251 \text{ MHz}. \tag{15.12}$$

The resolving power can be calculated from Eq. 15.1, or obtained directly from Eq. 15.6. We have

$$\mathcal{R} = 2.18 \times 10^6. \tag{15.13}$$

Problem 15.2. *What is the contrast factor of the FPI described in Problem 15.1? What would be the theoretical reduction in the intensity of the background, relative to the peaks, if the same FPI is used in a triple-pass configuration?*

From Eq. 15.7, the contrast factor of the FPI is

$$C = 361. \tag{15.14}$$

In a triple-pass configuration, the contrast factor would be

$$C^3 = 4.70 \times 10^7, \tag{15.15}$$

and the intensity of the background would be reduced by a factor

$$C^2 = 1.3 \times 10^5. \tag{15.16}$$

Problem 15.3. *A scanning confocal FPI (spectrum analyzer) is to be set up to study the longitudinal modes of an Ar^+ laser ($\lambda = 514$ nm). If the effective width of the gain profile of the laser medium is 6.5 GHz, what is the optimum radius of curvature of the mirrors in the FPI? If the laser cavity has a length of 860 mm, what should be the reflectance of the mirror coatings to resolve the individual modes?*

To avoid overlap of successive orders, the free spectral range (FSR) of the FPI should be greater than the bandwidth of the spectrum that is being studied. Since the effective width of the gain profile is 6.5 GHz, we can set the FSR at 7.0 GHz.

With an air-spaced confocal FPI, the optical path difference between successive reflected rays is four times the separation of the mirrors (which is equal to r, their radius of curvature), so that the FSR is given by the relation

$$\text{FSR}_\nu = c/4r. \tag{15.17}$$

Accordingly, we need mirrors with a radius of curvature

$$r = c/4\,\text{FSR}_\nu = 10.71 \text{ mm}. \tag{15.18}$$

The frequency difference between successive longitudinal modes of the laser (see Eq. 6.1) is equal to $c/2L$, where L is the length of the laser cavity. For a cavity length of 860 mm, the frequency difference between adjacent modes is, therefore, 173.8 MHz.

The width (FWHM) of the peaks of the FPI used as a spectrum analyzer must be significantly less than the separation of the modes to resolve them clearly. If we aim for a value of the FWHM of (say) $\Delta \nu_W = 35$ MHz, which is about 0.2 of the separation of the modes, we need a finesse

$$F = \text{FSR}_\nu / \Delta \nu_W = 200. \tag{15.19}$$

Since successive rays undergo two reflections at each mirror, the reflectance required for this finesse is the square root of that given by Eq. 5.6. We, therefore, need a reflectance

$$R = 0.993. \tag{15.20}$$

Problem 15.4. *The thinnest plate in a birefringent filter is made of quartz ($n_e - n_o = 0.009027$ at $\lambda = 656.3$ nm) and has a thickness $d = 0.9$ mm. An interference filter is used to isolate the transmission peak at this wavelength. How would you choose a suitable interference filter?*

The separation of adjacent peaks of the birefringent filter (which corresponds to its free spectral range) can be obtained from Eq. 15.8. We have

$$\text{FSR}_\nu = 1/\tau = c/(n_o - n_e)d. \tag{15.21}$$

Since $\nu\lambda = c$, the corresponding wavelength difference is

$$\begin{aligned}\text{FSR}_\lambda &= (\lambda^2/c)\text{FSR}_\nu \\ &= \lambda^2/(n_o - n_e)d \\ &= 53.02 \text{ nm}. \end{aligned} \tag{15.22}$$

To isolate the peak at $\lambda = 656.3$ nm, we need an interference filter with peak transmittance at this wavelength. The transmittance of this interference filter should drop to a negligible level for wavelengths separated from the peak by half the FSR of the birefringent filter, that is, by ± 26 nm. From available catalogs, we see that this requirement can be met by a three-cavity interference filter with a nominal pass band (FWHM) of 10 nm.

Problem 15.5. *The fringe counts obtained in a dynamic wavelength metre with a dye laser and a frequency-stabilized He–Ne laser, as the end reflector is moved over a distance of approximately 500 mm, are 1,621,207 and 1,564,426, respectively. The frequency of the reference He–Ne laser, which is locked to an absorption line of $^{127}I_2$, is 473,612,215 MHz. If the refractive index of air is 1.0002712, what is the wavelength of the dye laser in air?*

Since the speed of light (in a vacuum) is 299,792,458 m/sec, the wavelength of the reference laser in a vacuum is

$$\lambda_2(\text{vac}) = 0.632991398 \ \mu\text{m}. \tag{15.23}$$

The vacuum wavelength of the dye laser is obtained by multiplying the wavelength of the reference laser by the ratio of the fringe counts. We have

$$\lambda_1(\text{vac}) = (1{,}564{,}426/1{,}621{,}207)\lambda_2(\text{vac})$$
$$= 0.6108216\ \mu\text{m}. \tag{15.24}$$

Accordingly, the wavelength of the dye laser in air is

$$\lambda_1(\text{air}) = 0.6108216/1.0002712$$
$$= 0.6106560\ \mu\text{m}. \tag{15.25}$$

FURTHER READING

For more information, see

1. G. Hernandez, *Fabry–Perot Interferometers*, Cambridge University Press, Cambridge, UK (1986).
2. J. M. Vaughan, *The Fabry–Perot Interferometer*, Adam Hilger, Bristol (1989).
3. H. A. MacLeod, *Thin-Film Optical Filters*, Institute of Physics Publishing, Bristol (2001).
4. P. Hariharan, *Optical Interferometry*, Academic Press, San Diego (2003).

16

Fourier Transform Spectroscopy

In Fourier transform spectroscopy (FTS), the intensity at a point in the interference pattern is recorded, as the optical path difference in the interferometer is varied, to yield what is known as the interference function. The variable part of the interference function (called the interferogram), on Fourier transformation (see Appendix II), yields the spectrum. FTS is now used widely in the infrared region because of the improved signal-to-noise (S/N) ratio, high throughput, and high resolution possible with it.

Some of the topics that we will discuss in this chapter are

- The multiplex advantage
- The theory of FTS
- Practical aspects of FTS
- Computation of the spectrum
- Applications of FTS

16.1 THE MULTIPLEX ADVANTAGE

When a spectrometer is operated in the scanning mode, the total scanning time T is divided between, say, m elements of the spectrum, so that each element of the spectrum is observed only for a time (T/m). In the far infrared region, the energy of individual photons is low, and the main source of noise is the detector. Since the noise power is independent of the signal, the signal-to-noise (S/N) ratio is reduced by a factor $m^{1/2}$.

However, if the optical path difference in an interferometer is varied linearly with time, each element of the spectrum gives rise to an output that is modulated at a frequency inversely proportional to its wavelength. It is then possible to record all these signals simultaneously (or, in other words, to multiplex them) and decode them later to obtain the spectrum. Since each spectral element is now recorded over the full scan time T, an improvement in the S/N ratio by a factor of $m^{1/2}$ over a conventional scanning instrument (the multiplex advantage) is obtained.

16.2 THEORY

Consider a two-beam interferometer illuminated with a collimated beam. The beam emerging from the interferometer is focused on a detector. With monochromatic light, the output from the detector can be written as a function of the delay $\tau = p/c$, where p is the optical path difference, in the form

$$G(\tau) = g(\nu)(1 + \cos 2\pi \nu \tau)$$

$$= g(\nu) + g(\nu) \cos 2\pi \nu \tau, \qquad (16.1)$$

where

$$g(\nu) = L(\nu)T(\nu)D(\nu) \qquad (16.2)$$

is the product of three spectral distributions: the radiation studied, $L(\nu)$; the transmittance of the spectroscope, $T(\nu)$; and the detector sensitivity, $D(\nu)$.

With a source having a large spectral bandwidth, we have to integrate Eq. 16.1 over the entire range of frequencies, and the output of the detector is

$$G(\tau) = \int_0^\infty g(\nu)\, d\nu + \int_0^\infty g(\nu) \cos 2\pi \nu \tau \, d(\nu). \qquad (16.3)$$

Since the first term on the right-hand side of Eq. 16.3 is a constant, the variable part of the output, which constitutes the interferogram, is

$$F(\tau) = \int_0^\infty g(\nu) \cos 2\pi \nu \tau \, d(\nu). \qquad (16.4)$$

Since all the spectral components are in phase when $\tau = 0$, the interferogram initially exhibits large fluctuations as τ is increased from zero (see Figure 16.1), but the amplitude of these fluctuations drops off rapidly. Fourier inversion of Eq. 16.4 then gives

$$g(\nu) = 4 \int_0^\infty F(\tau) \cos 2\pi \nu \tau \, d(\tau). \qquad (16.5)$$

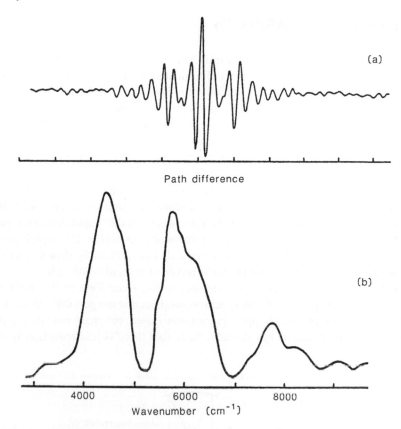

Figure 16.1. (a) Interferogram, and (b) spectrum, obtained with the 3.5 magnitude star, α Herculis (P. Fellgett, *J. Phys. Radium* **19**, 237–240, 1958).

The resolution that can be obtained is limited by the maximum value of the delay. If the interferogram is truncated at $\pm\tau_m$, the resolution limit $\Delta\nu$ is

$$\Delta\nu = 1/2\tau_m. \tag{16.6}$$

Truncation of the interferogram is undesirable, since it produces side lobes that could be mistaken for other weak spectral lines. These side lobes can be eliminated, at some loss in resolution, by multiplying the interferogram with a weighting function which progressively reduces the contribution of greater delays. This process is known as apodization.

16.3 PRACTICAL ASPECTS

The main component of a Fourier transform spectrometer is, as shown in Figure 16.2, a Michelson interferometer illuminated with an approximately collimated beam. In the near infrared region, thin films of Ge or Si on CaF_2 or KBr plates are used as beam splitters, while a thin film of Mylar, or a wire mesh, can be used in the far infrared. The slide carrying the moving mirror must be of very high quality to avoid tilting; this problem can be minimized by replacing the mirrors with cat's-eye reflectors consisting of concave mirrors with small plane mirrors placed at their foci.

Two approaches to the movement of the mirror (scanning) have been followed. In periodic generation, the mirror is moved repeatedly over the desired scanning range at a rate sufficiently rapid that the fluctuations of the output due to the passage of the interference fringes occur at a frequency permitting AC amplification. In aperiodic generation, the mirror is moved only once, relatively slowly, over the scanning range, and the detector output is recorded at regular intervals.

With aperiodic generation, it is necessary to use some form of flux modulation, so that AC amplification and synchronous detection are possible. Amplitude modulation (by means of a chopper) is commonly used, but phase modulation (by vibrating the fixed mirror) has the advantage that there is less reduction in the output.

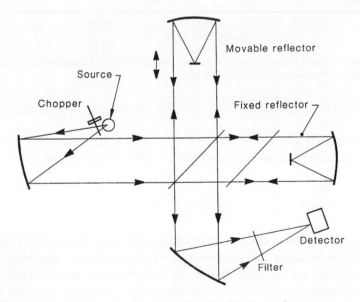

Figure 16.2. Michelson interferometer adapted for Fourier transform spectroscopy.

16.4 COMPUTATION OF THE SPECTRUM

To compute the spectrum from the interferogram, it is necessary to sample the interferogram at a number of equally spaced points. To avoid ambiguities, the increment in the optical path difference between samples must satisfy the condition

$$\Delta p < \lambda_{min}/2, \tag{16.7}$$

where λ_{min} is the shortest wavelength in the spectrum being recorded.

The computation of the spectrum has been greatly speeded up by the use of the fast Fourier transform (FFT) algorithm. The total number of operations involved in computing a Fourier transform by conventional routines is approximately $2M^2$, where M is the number of points at which the interferogram is sampled. With the FFT algorithm, the number of operations is reduced to $3M \log_2 M$, so that the computation of complex spectra becomes feasible.

16.5 APPLICATIONS

Fourier transform spectroscopy has found many applications in the far infrared region; they include studies of emission spectra, absorption spectra, chemiluminescence, and the kinetics of chemical reactions. In addition, because of their high etendue, Fourier transform spectrometers can be used to record high-resolution spectra from very faint sources, such as planetary atmospheres and the night sky.

16.6 SUMMARY

- FTS is most useful in the far infrared region, where it gives a much better S/N ratio than wavelength scanning.
- FTS can be used to record high-resolution spectra of very faint sources.

16.7 PROBLEMS

Problem 16.1. *A Fourier transform spectrometer is required to record spectra in the far infrared region over the wavenumber range from 40 cm^{-1} to 250 cm^{-1} (note: wavenumber $\sigma = 1/\lambda$). What would be the minimum optical path difference (OPD) over which the interferogram would have to be recorded to obtain a resolution of 0.13 cm^{-1}?*

If we divide both sides of Eq. 16.6, which gives the resolution limit in terms of the delay, by the speed of light (c), we have

$$\Delta\sigma = 1/p, \tag{16.8}$$

where p is the OPD required for a resolution $\Delta\sigma$ (in wavenumbers). Accordingly, for a resolution of 0.13 cm^{-1}, we would require the interferogram to be recorded over an OPD of 7.69 cm. In practice, the interferogram would have to be apodized to eliminate side lobes; this would reduce the resolution by a factor of 2. We would, therefore, need to record the interferogram over an OPD of $\pm$15.38 cm.

Problem 16.2. *In Problem 16.1, what is the number of points at which the interferogram should be sampled? What is the theoretical gain in the S/N ratio over a conventional scanning instrument having the same resolution?*

Equation 16.7, which specifies the maximum increment in the OPD between samples that avoids ambiguities, can be rewritten in the form

$$\Delta p < 1/2\sigma_{max}, \tag{16.9}$$

which yields an upper limit for the sampling interval of 20 μm.

Since we need to take samples at this interval up to the maximum value of the OPD on both sides of the origin, the interferogram must be sampled at a minimum of 15,380 points. However, the FFT algorithm requires 2^N data points, where N is an integer. We would therefore sample the interferogram at $2^{14} = 16,384$ points.

The theoretical improvement in the S/N ratio over a conventional scanning instrument, for the same number of data points, would be $(16,384)^{1/2} = 128$.

Problem 16.3. *For the number of sampling points used in Problem 16.2, how many operations would be involved in computing the spectrum from the interferogram (a) by conventional routines, and (b) using the FFT algorithm? What would be the reduction in the computation time obtained by using the FFT algorithm?*

The number of operations required in the conventional method of computing the Fourier transform would be

$$2 \times 16,384^2 = 2.684 \times 10^8. \tag{16.10}$$

On the other hand, the number of operations required with the FFT algorithm would be

$$3 \times 16,384 \times \log_2 16,384 = 688,128. \tag{16.11}$$

The computation time would, therefore, be reduced by a factor of 780.

FURTHER READING

For more information, see

1. R. J. Bell, *Introductory Fourier Transform Spectroscopy*, Academic Press, New York (1972).
2. J. Chamberlain, *The Principles of Interferometric Spectroscopy*, Wiley, Chichester (1979).
3. S. P. Davis, M. C. Abrams, and J. W. Brault, *Fourier Transform Spectrometry*, Academic Press, San Diego (2001).

17

Interference with Single Photons

Young's classical experiment has always been regarded as a conclusive demonstration of the wavelike nature of light. However, at very low light levels, photodetectors register distinct events corresponding to the annihilation of individual photons. It follows that light cannot be either a particle or a wave, but exhibits the characteristics of both. In this chapter we will discuss

- Interference—the quantum picture
- Single-photon states
- Interference with single-photon states
- Interference with independent sources
- Fourth-order interference

17.1 INTERFERENCE—THE QUANTUM PICTURE

We can say that we are in the "single-photon" regime when, with a perfectly efficient detector, the mean time interval between the detection of successive photons is much greater than the time taken for light to travel through the system. It then follows that, if interference involves the interaction of two photons, interference effects should disappear when, at a time, only a single photon is in the interferometer. However, experiments at extremely low light levels have confirmed that the quality of an interference pattern does not depend on the intensity.

In the quantum picture, interference involves mixing two fields. The result is a two-mode coherent state. In the weak field limit, this state represents the interference of a photon with itself, and can be interpreted as a sum over histories, as outlined below.

A photon can take either of two paths from the source to the detector. Associated with each path is a complex probability amplitude a_i $(i = 1, 2)$, which we can write explicitly as

$$a_1 = |a_1| \exp(i\phi_1),$$
$$a_2 = |a_2| \exp(i\phi_2). \tag{17.1}$$

The absolute square of each of these complex probability amplitudes represents the probability of the photon taking that path and corresponds to the intensity at the detector due to that path acting in isolation.

The probability of detecting a photon, that is to say, the intensity at the detector, when both paths are open, is obtained by summing the probability amplitudes for the two paths and taking the square of the modulus of the sum, so that we have

$$I = |a_1 + a_2|^2, \tag{17.2}$$

which is the familiar equation for two-beam interference.

17.2 SINGLE-PHOTON STATES

With a thermal source, or even a single-mode laser source, photons are more likely to arrive at a photodetector very close together, than far apart in time. This phenomenon is known as photon bunching. Such sources cannot, therefore, generate a single-photon state.

An approximation to a single-photon state can be produced by generating a pair of photons. The detection of one photon then signals the presence of a second photon, whose frequency and direction of propagation are related to those of the first photon. The second photon can then be regarded as being in a one-photon Fock state.

One way to generate such a pair of photons is by an atomic cascade, where a calcium atom emits two photons in rapid succession. If atoms of calcium are excited to the 6^1S_0 level, they return to the ground state by a two-step process in which they emit, in rapid succession, two photons with wavelengths of 551.3 nm and 422.7 nm, respectively.

However, a better method is parametric down-conversion, in which a single UV (*pump*) photon spontaneously decays in a crystal with a $\chi^{(2)}$ nonlinearity into a *signal* photon and an *idler* photon, with wavelengths close to twice the wavelength of the UV photon, and polarizations orthogonal to that of the UV photon. Phase matching between the UV beam and the down-converted beams, to maximize the output, is achieved by using a birefringent crystal.

UV pump beam $(\lambda = 351.1 \text{ nm})$

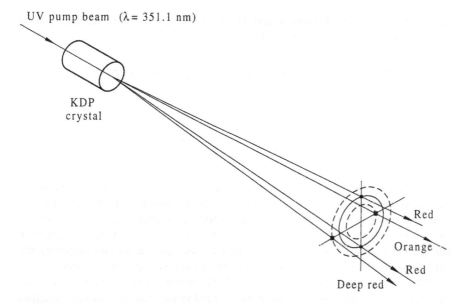

KDP
crystal

Red

Orange

Red

Deep red

Figure 17.1. Generation of photon pairs by parametric down-conversion of UV photons in a nonlinear crystal.

Since energy is conserved, we have

$$\hbar\omega_0 = \hbar\omega_1 + \hbar\omega_2, \tag{17.3}$$

where $\hbar\omega_0$ is the energy of the UV photon, and $\hbar\omega_1$ and $\hbar\omega_2$ are the energies of the two down-converted photons. Similarly, since momentum is conserved, we have

$$\mathbf{k}_0 = \mathbf{k}_1 + \mathbf{k}_2, \tag{17.4}$$

where $\mathbf{k}_0$ is the momentum of the UV photon, and $\mathbf{k}_1$ and $\mathbf{k}_2$ are the momenta of the down-converted photons. Accordingly, the photons in each pair are emitted on opposite sides of two cones, whose axis is the UV beam, and produce, as shown in Figure 17.1, a set of rainbow-colored rings.

The down-converted photons carry information on the phase of the pump and are in an entangled state. As a result, the state of the idler photon is governed by any measurements made on the signal photon, and vice versa.

Typically, the UV beam from an Ar^+ laser $(\lambda = 351.1 \text{ nm})$ and a potassium dihydrogen phosphate (KDP) crystal can be used to generate pairs of photons with wavelengths around 746 and 659 nm, leaving the crystal at angles of approximately $\pm 1.5°$ to the UV pump beam. Higher down-conversion efficiency can be obtained with beta barium borate (BBO) crystals.[1]

17.3 INTERFERENCE WITH SINGLE-PHOTON STATES

With a single-photon state, quantum mechanics predicts a perfect anticorrelation between the counts at the two outputs from a beam splitter. This behavior is very different from that with a thermal source, where the correlation between the two outputs is positive, or with the coherent field from a laser, where there is no correlation between the two outputs.

Interference effects produced by single-photon states were first studied using an atomic cascade. As shown in Figure 17.2(a), the arrival of the first photon of a pair (frequency ν_1) at the detector D_0 triggered a gate, enabling the two detectors D_1 and D_2 for a very short time τ, so as to maximize the probability of detecting the second photon (frequency ν_2) emitted by the same atom and minimize the probability of detecting a photon emitted by any other atom in the source.

While a classical wave would be divided between the two output ports of the beam splitter, a single photon cannot be divided in this fashion. We can therefore expect an anticorrelation between the counts on the two sides of the beam splitter at D_1 and D_2, measured by a parameter

$$\mathcal{A} = N_{012}N_0/N_{01}N_{02}, \tag{17.5}$$

where N_{012} is the rate of triple coincidences between the detectors D_0, D_1, and D_2; N_{01} and N_{02} are, respectively, the rate of coincidences between D_0 and D_1, and D_0 and D_2; and N_0 is the rate of counts at D_0.

For a classical wave, $\mathcal{A} \geqslant 1$. On the other hand, the indivisibility of the photon should lead to arbitrarily small values of $\mathcal{A}$. With a gate time $\tau = 9$ ns, the number of coincidences observed was only 0.18 of that expected from classical theory, confirming the presence of a single-photon state.

The same source was then used in the optical arrangement shown in Figure 17.2(b), with the detectors D_1 and D_2 receiving the two outputs from a Mach–Zehnder interferometer. The interferometer was initially adjusted without the gating system, and interference fringes with a visibility $\mathcal{V} > 0.98$ were obtained. The gate was then turned on, and the optical path difference was varied around zero in 256 steps, each of $\lambda/50$, with a counting time of 1 sec at each step. Analysis of the data obtained showed that, even with the gate operating, interference fringes with a visibility $\mathcal{V} > 0.98$ were obtained at both outputs.

This result confirms the quantum picture of interference, namely that a photon can be regarded as interfering with itself.

---

[1] See D. Dehlinger and M. W. Mitchell, "Entangled photon apparatus for the undergraduate laboratory," *Am. J. Phys.* **70**, 898–902 (2002).

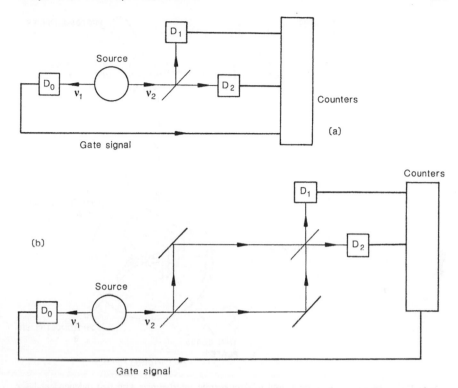

Figure 17.2. Experimental arrangement used (a) to detect single-photon states, and (b) to demonstrate interference with single-photon states (P. Grangier, G. Roger, and A. Aspect, *Europhys. Lett.* **1**, 173–179, 1986).

17.4 INTERFERENCE WITH INDEPENDENT SOURCES

Pfleegor and Mandel were the first to show that light beams from two independent lasers can produce interference fringes, even when the light intensity is so low that the mean time between photons is long enough that there is a high probability that one photon is absorbed at the detector before the next photon is emitted by either of the two sources. As shown in Figure 17.3, the beams from two He–Ne lasers, operating at the same wavelength, were superimposed at a small angle to produce interference fringes on the edges of a stack of glass plates whose thickness was equal to half the fringe spacing. Two photomultipliers received the light from alternate plates. To minimize effects due to movements of the fringes, an additional photodetector was used to detect beats between the beams, and measurements were restricted to 20 μs intervals, corresponding to periods during which the frequency difference between the two laser beams was less than 30 kHz.

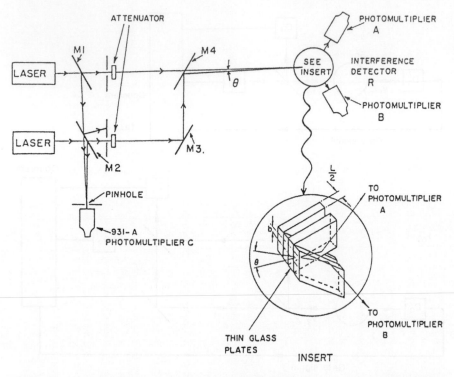

Figure 17.3. Experimental system used to demonstrate interference with two independent laser sources at very low light levels (R. L. Pfleegor and L. Mandel, *Phys. Rev.* **159**, 1084–1088, 1967).

While the positions of the fringe maxima may vary from measurement to measurement, there should always be a negative correlation between the number of photons registered in the two channels, which should be a maximum when the fringe spacing l is equal to L, the thickness of a pair of plates. Figure 17.4 shows the variation in the degree of correlation of the two counts with the ratio L/l, together with the theoretical curves for $N = 2$ and $N = 3$, where N is the number of plates in the detector array.

The correlation that was observed confirmed that interference effects were associated with the detection of each photon. However, since observations could be made only over very short time intervals, during which only a small number of photons were detected, the precision of the experiment was limited.

More precise observations of interference effects with two sources at the single-photon level have been made by using the low-frequency beat between two laser modes obtained by applying a transverse magnetic field to an He–Ne laser oscillating in two longitudinal modes. The two, orthogonally polarized, Zeeman-split components can be regarded as equivalent to beams from two independent

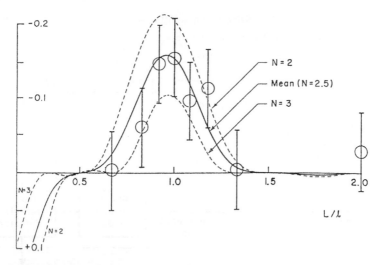

Figure 17.4. Experimental results for the normalized correlation coefficient (solid line) and theoretical curves for $N = 2$ and $N = 3$ (broken lines) (R. L. Pfleegor and L. Mandel, *Phys. Rev.* **159**, 1084–1088, 1967).

lasers, because there is no coherence between the two upper states for the lasing transitions.

As shown in Figure 17.5, the beat frequency was stabilized by mixing the two orthogonally polarized components in the back beam at a monitor photodiode, and using the output to control the length of the laser cavity by a heating coil. Neutral-density filters were used to reduce the output intensity in known steps over a range of $10^8 : 1$, and the attenuated beams, after passing through a polarizing beam-splitter, were incident on a photodiode. The signal from the photodiode was taken to a homodyne detector, which also received a reference signal from the monitor photodiode. Because the variations in the frequency of the beat signal were tracked by the reference signal, measurements could be made with integrating times up to 100 seconds, ensuring a good signal-to-noise ratio even at the lowest light levels.

Observations were made with the photodiode at a distance of 0.2 m from the laser, as the incident power was varied from 1.0 μW down to 4.8 pW. At the lowest power level, the probability for the presence of more than one photon in the apparatus at any time, relative to that for a single photon, was less than 0.005.

The output from the homodyne detector, when plotted as a function of the power incident on the photodiode (see Figure 17.6), showed no significant deviations from a straight line, confirming that the interference effects observed remained unchanged down to the lowest power levels.

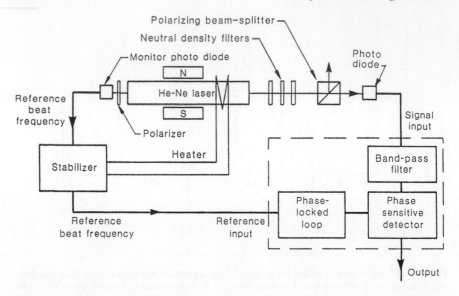

Figure 17.5. Experimental arrangement used to measure the amplitude of beats between two laser modes at the single-photon level (P. Hariharan, N. Brown, and B. C. Sanders, *J. Mod. Opt.* **40**, 113–122, 1993).

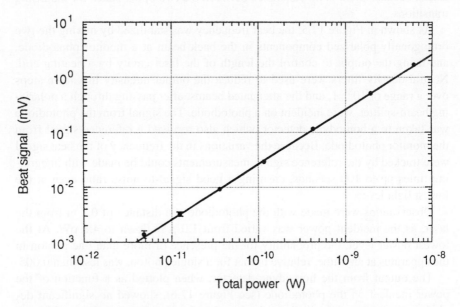

Figure 17.6. Output signal from the homodyne detector as a function of the total power in the laser beams (P. Hariharan, N. Brown, and B. C. Sanders, *J. Mod. Opt.* **40**, 113–122, 1993).

17.5 FOURTH-ORDER INTERFERENCE

Correlated photon pairs produced by parametric down-conversion exhibit effects which cannot be observed with classical light sources.

If the two photons are the inputs to the two ports of a Mach–Zehnder interferometer, the photon count rates at the two output ports remain unchanged when the optical path difference is varied. However, the rate of coincidences exhibits a sinusoidal variation (interference fringes). This behavior can be attributed to the fact that when two photons in an entangled state enter a beam splitter simultaneously at the two input ports, they always emerge together, at one or the other of the output ports. The resulting fourth-order interference fringes are due to the interference of photon pairs rather than single photons.

Fourth-order interference effects can also be observed if, as shown in Figure 17.7, the two photons are sent into two separate interferometers, which are adjusted so that the difference in the lengths of the optical paths in each interferometer is more than the coherence length of the individual photons, but is very nearly the same in both interferometers. Under these conditions, the photon count rates at the output from each of the two interferometers show no variation with the optical path difference. However, measurements of the rate of coincidences, as a function of the difference in the imbalances, show variations with a period corresponding to the wavelength of the pump beam.

This result constitutes a violation of Bell's inequality and contradicts the Einstein–Podolsky–Rosen postulate.

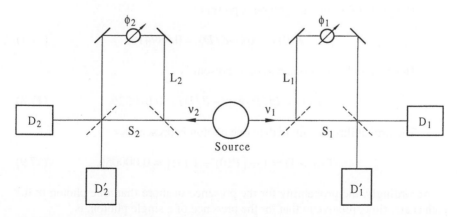

Figure 17.7. Experiment using fourth-order interference to demonstrate a violation of Bell's inequality (J. D. Franson, *Phys. Rev. Lett.* **62**, 2205–2208, 1989).

17.6 SUMMARY

- The "photon picture" can be applied to interference phenomena, provided we apply Dirac's dictum that "... a photon interferes only with itself."
- We can then regard optical interference as due to the existence of indistinguishable paths.
- The addition of the complex probability amplitudes associated with each path, and the evaluation of the squared modulus of this sum, yields the probability of detection of a photon.

17.7 PROBLEMS

Problem 17.1. *The beam from an He–Ne laser ($\lambda = 633$ nm) is attenuated by a set of neutral-density filters so that the power in the output beam is 5 pW (5×10^{-12} W). What is the mean distance between photons? If a photodiode is at a distance of 0.2 m from the laser, what is the probability for the presence of more than one photon in the path at any time, relative to that for a single photon?*

At a wavelength of 633 nm, the energy of a single photon is 3.14×10^{-19} J. A power level of 5 pW, therefore, corresponds to a mean photon flux $N = 1.59 \times 10^7$ photons/second. Accordingly, the mean distance between photons is

$$D = c/N = 3 \times 10^8 / 1.59 \times 10^7 = 18.8 \text{ m}. \qquad (17.6)$$

Since the photons in a laser beam exhibit a Poisson distribution, it follows that, in an optical path with a length $d = 0.2$ m:

1. The probability that no photon is present is

$$P(0) = \exp(-d/D) = 0.989418. \qquad (17.7)$$

2. The probability that one photon is present is

$$P(1) = (d/D)\exp(-d/D) = 0.010526. \qquad (17.8)$$

3. The probability that more than one photon is present is

$$P(n > 1) = 1 - \left[P(0) + P(1)\right] = 0.000056. \qquad (17.9)$$

Accordingly, the probability for the presence of more than one photon in the path at any time, relative to that for the presence of a single photon, is

$$P(n > 1)/P(1) = 0.0053. \qquad (17.10)$$

FURTHER READING

For more information, see

1. R. W. Boyd, *Nonlinear Optics*, Academic Press, Boston (1992).
2. R. P. Feynman, R. B. Leighton, and M. Sands, *Lectures on Physics*, Vol. 3, Addison-Wesley, London (1963).
3. P. Hariharan and B. C. Sanders, *Quantum Effects in Optical Interference*, in Progress in Optics, Vol. XXXVI, Ed. E. Wolf, Elsevier, Amsterdam (1996), pp. 49–128.
4. L. Mandel and E. Wolf, *Optical Coherence and Quantum Optics*, Cambridge University Press, Cambridge (1995).

18

Building an Interferometer

Choosing an interferometer for any application often involves deciding whether to buy or build one.

Interferometers are available off the shelf for many applications. They include

- Measurements of length
- Optical testing
- Interference microscopy
- Laser–Doppler interferometry
- Interferometric sensors
- Interference spectroscopy
- Fourier-transform spectroscopy

A considerable amount of information on suppliers of such instruments is available from technical journals dealing with these fields.

However, there are many experiments that only require a comparatively simple optical system; there are also specialized problems that cannot be handled with commercially available instruments. In such cases, a breadboard setup is cheaper and more flexible, and often not as difficult to put together as commonly imagined.

A layout for a breadboard setup for a Michelson/Twyman–Green interferometer is shown in Figure 18.1.

For those adventurous souls who would like to have fun building an interferometer, a wide range of assemblies and components are available that make it possible to build quite sophisticated systems for specific purposes. Some of these items are

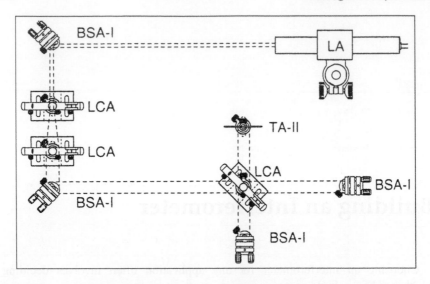

Figure 18.1. Schematic of a breadboard setup for a Michelson/Twyman–Green interferometer (from *Projects in Optics: Applications Workbook*, ©Newport Corporation, reproduced with permission).

- Lasers
- Optical tables
- Optical mounting hardware
- Lenses, mirrors, and beam splitters
- Polarizers and retarders
- Photodetectors
- Image processing hardware and software

An excellent source of information on suppliers of such items is

- Physics Today Buyers Guide
  www.physicstoday.org/guide/

and another is:

- OSA's Online Product Guide
  www.osa.org/Product_Guide/

while information on a wide range of optical components are available from:

- Edmund Optics
  www.edmundoptics.com

Detailed drawings of a large number of optical assemblies are available on the websites of most suppliers. Drawings of appropriate assemblies can be selected and put together to design a completely new setup rapidly and with minimum effort.

FURTHER READING

Those who choose to follow the do-it-yourself path may find some useful hints in

1. C. H. Palmer, *Optics, Experiments and Demonstrations*, Johns Hopkins Press, Baltimore (1962).
2. *Projects in Optics: Applications Workbook*, Newport Corporation, Fountain Valley, California.
3. J. T. McCrickerd, *Projects in Holography*, Newport Corporation, Fountain Valley, California (1982).
4. R. S. Sirohi, *A Course of Experiments with the He–Ne Laser*, John Wiley, New York (1986).
5. K. Izuka, *Engineering Optics*, Springer-Verlag, Berlin (1987).
6. J. Strong, *Procedures in Applied Optics*, Marcel Dekker, New York (1989).

A

Monochromatic Light Waves

A.1 COMPLEX REPRESENTATION

The time-varying electric field at any point in a beam of light (amplitude a, frequency ν, wavelength λ) propagating in the z direction can be written as

$$E(z,t) = a\cos\left[2\pi(\nu t - z/\lambda)\right]$$
$$= a\cos(\omega t - kz), \tag{A.1}$$

where $\omega = 2\pi\nu$, and $k = 2\pi/\lambda$. However, many mathematical operations can be handled more simply with a complex exponential representation. Accordingly, it is convenient to write Eq. A.1 in the form

$$E(z,t) = \mathrm{Re}\left\{a\exp(i\omega t - kz)\right\}, \tag{A.2}$$

where $\mathrm{Re}\{\ \}$ is the real part of the expression within the braces, and $i = (-1)^{1/2}$. The right-hand side can then be separated into a spatially varying factor and a time-varying factor and rewritten as

$$E(z,t) = \mathrm{Re}\left\{a\exp(-ikz)\exp(i\omega t)\right\},$$
$$= \mathrm{Re}\left\{a\exp(-i\phi)\exp(i\omega t)\right\},$$
$$= \mathrm{Re}\left\{A\exp(i\omega t)\right\}, \tag{A.3}$$

where $\phi = 2\pi z/\lambda$, and $A = a\exp(-i\phi)$ is known as the complex amplitude.

169

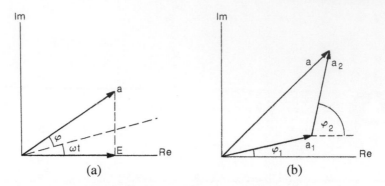

Figure A.1. Phasor representation of light waves.

On an Argand diagram, as shown in Figure A.1(a), this representation corresponds to the projection on the real axis of a vector of length a, initially making an angle ϕ with the real axis and rotating counterclockwise at a rate ω (a phasor).

An advantage of such a representation is that the resultant complex amplitude due to two or more waves, having the same frequency and traveling in the same direction, can be obtained by simple vector addition. In this case, since the phasors are rotating at the same rate, we can neglect the time factor and the resultant complex amplitude is given, as shown in Figure A.1(b), by the sum of the complex amplitudes of the component waves.

A.2 OPTICAL INTENSITY

The energy that crosses unit area normal to the direction of propagation of the beam, in unit time, is proportional to the time average of the square of the electric field

$$\langle E^2 \rangle = T \overset{\lim}{\to} \infty \frac{1}{2T} \int_{-T}^{T} E^2 \, dt$$

$$= a^2/2. \tag{A.4}$$

Conventionally, the factor of $1/2$ is ignored, and the optical intensity is defined as

$$I = a^2$$

$$= AA^* = |A|^2, \tag{A.5}$$

where $A^* = a \exp(i\phi)$ is the complex conjugate of A.

B

Phase Shifts on Reflection

Consider a light wave of unit amplitude incident, as shown in Figure B.1(a), on the interface between two transparent media. This incident wave gives rise to a reflected wave (amplitude r) and a transmitted wave (amplitude t).

If the direction of the reflected wave is reversed, as shown in Figure B.1(b), it will give rise to a reflected component with an amplitude r^2 and a transmitted component with an amplitude rt. Similarly, if the transmitted wave is reversed, it will give rise to a reflected component with an amplitude tr' and a transmitted component with an amplitude tt' (where r' and t' are, respectively, the reflectance and transmittance, for amplitude, for a ray incident on the interface from below).

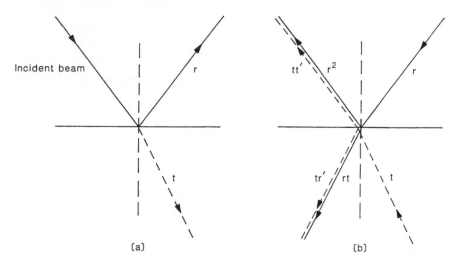

Figure B.1.   Phase shifts on reflection at the interface between two transparent media.

If there are no losses at the interface, it follows that

$$r^2 + tt' = 1,$$
$$rt + tr' = 0, \tag{B.1}$$

so that

$$r' = -r. \tag{B.2}$$

Equation B.2 shows that the phase shifts for reflection at the two sides of the interface differ by π, corresponding to the introduction of an additional optical path in one beam of $\lambda/2$.

It should be noted that this simple relation only applies to the interface between two transparent media; the phase changes on reflection at a metal film are more complicated.

C

Diffraction

The edge of the shadow of an object illuminated by a point source is not sharply defined, as one might expect from geometrical considerations, but exhibits bright and dark regions. The deviation of light waves from rectilinear propagation, in this manner, is known as diffraction. Effects due to diffraction are observed whenever a beam of light is restricted by an aperture or an edge.

Diffraction can be explained in terms of Huygens' construction. Each point on the unobstructed part of the wavefront can be considered a source of secondary wavelets. All these secondary wavelets combine to produce the new wavefront that is propagated beyond the obstruction. The complex amplitude of the field, at any point beyond the obstruction, can be obtained by summing the complex amplitudes due to these secondary sources and, in the most general case, is given by the Fresnel–Kirchhoff integral.

A special case, of particular interest, is when the source and the plane of observation are at an infinite distance from the diffracting aperture. We then have what is known as Fraunhofer diffraction. This situation commonly arises when the object is illuminated with a collimated beam, and the diffraction pattern is viewed in the focal plane of a lens. Fraunhofer diffraction also occurs when the lateral dimensions of the object (x, y) are small enough, compared to the distances to the source and the plane of observation, to satisfy the far-field condition

$$z \gg (x^2 + y^2)/\lambda. \tag{C.1}$$

The field distribution in the Fraunhofer diffraction pattern can be calculated quite easily, since it is the two-dimensional Fourier transform (see Appendix H) of the field distribution across the diffracting aperture.

A typical example is a small circular pinhole illuminated by a distant point source. In this case, as shown in Figure C.1, the Fraunhofer diffraction pattern consists of a central bright spot (the Airy disk) surrounded by concentric dark and

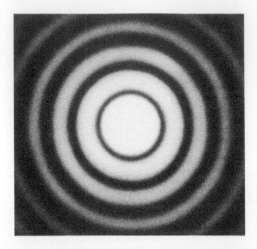

Figure C.1.   Diffraction pattern produced by a small circular hole.

bright rings. The angle 2θ subtended by the first dark ring at the pinhole is given by the relation

$$\sin \theta = 1.22\lambda/d, \tag{C.2}$$

where d is the diameter of the pinhole. For a pinhole with a diameter of 0.1 mm illuminated by an He–Ne laser ($\lambda = 0.633$ μm), the diameter of the first dark ring in the diffraction pattern observed on a screen at a distance of 1 metre would be 15.5 mm.

C.1 DIFFRACTION GRATINGS

A transmission diffraction grating consists of a set of equally spaced, parallel grooves on a transparent substrate; a reflection grating consists of a similar set of grooves on a reflecting substrate.

Light incident on a transmission grating at an angle θ_i is diffracted, as shown in Figure C.2(a), at specific angles θ_m given by the relation

$$\Lambda(\sin \theta_m - \sin \theta_i) = m\lambda, \tag{C.3}$$

where Λ is the groove spacing, and m is an integer. With a reflection grating, as shown in Figure C.2(b),

$$\Lambda(\sin \theta_m + \sin \theta_i) = m\lambda. \tag{C.4}$$

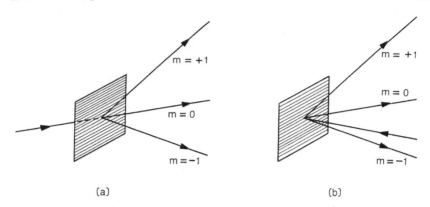

Figure C.2. Diffraction of light by (a) a transmission grating, and (b) a reflection grating.

Diffraction gratings can be blazed, by modifying the shape of the grooves, to diffract most of the incident light into a specified diffracted order.

D

Polarized Light

D.1 PRODUCTION OF POLARIZED LIGHT

With unpolarized light, the electric field vector does not have any preferred orientation and moves rapidly, through all possible orientations, in a random manner. Polarized light can be generated from unpolarized light by transmission through a sheet polarizer (or a polarizing prism) that selects the component of the field vector parallel to the principal axis of the polarizer. Light can also be polarized by reflection. If unpolarized light is incident, as shown in Figure D.1, on the interface between two transparent media with refractive indices n_1 and n_2, the component with the field vector parallel to the reflecting surface is preferentially reflected. At an angle of incidence θ_B (the Brewster angle) given by the condition

$$\tan\theta_B = n_2/n_1, \tag{D.1}$$

the reflected light is completely polarized, with its field vector parallel to the reflecting surface.

When polarized light is incident on a polarizer whose axis makes an angle θ with the field vector, the light emerging is linearly polarized in a direction parallel to the axis of the polarizer, but its intensity is

$$I = I_0 \cos^2\theta, \tag{D.2}$$

where I_0 is the intensity when $\theta = 0$.

D.2 QUARTER-WAVE AND HALF-WAVE PLATES

A birefringent material is characterized by two unique directions, perpendicular to each other, termed the fast and slow axes. The refractive index for light

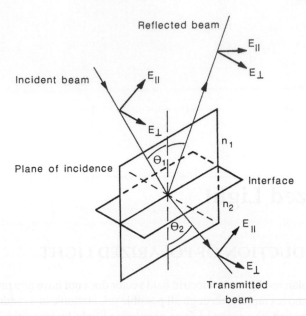

Figure D.1. Incident, transmitted, and reflected fields at the interface between two transparent media.

polarized parallel to the fast axis (n_f) is less than that for light polarized parallel to the slow axis (n_s). A beam of unpolarized light incident on such a material is resolved into two components that are polarized at right angles to each other. These two components travel at different speeds and, in general, are refracted at different angles.

A case of interest is a wave (wavelength λ), polarized at 45° with respect to these two axes, incident normally on a birefringent plate of thickness d. The two orthogonally polarized waves that emerge then have the same amplitude, but exhibit a phase difference

$$\Delta\phi = (2\pi/\lambda)(n_s - n_f)d$$
$$= (2\pi/\lambda)\delta, \tag{D.3}$$

where $\delta = (n_s - n_f)d$ is the optical path difference (the retardation) introduced between the two waves by the birefringent plate.

If the thickness of the plate is such that

$$\delta = (n_s - n_f)d = \lambda/4, \tag{D.4}$$

it is called a quarter-wave retarder, or a quarter-wave ($\lambda/4$) plate. The phase difference between the two waves when they emerge from the plate is

$$\Delta\phi = \pi/2, \tag{D.5}$$

and the projection of the field vector on the x, y plane traces out a circle. The light is then said to be circularly polarized. If the fast and slow axes of the plate are interchanged by rotating it through $90°$, light that is circularly polarized in the opposite sense is obtained.

If the thickness of the plate is doubled, we have a half-wave ($\lambda/2$) plate (a half-wave retarder), and the phase difference between the beams, when they emerge, is π. In this case, the light remains linearly polarized, but its plane of polarization is rotated through an angle 2θ, where θ is the angle between the fast axis of the half-wave plate and the incident field vector.

D.3 THE JONES CALCULUS

In the Jones calculus, the characteristics of a polarized beam are described by a two-element column vector. For a plane monochromatic light wave propagating along the z-axis, with components a_x and a_y along the x- and y-axes, the Jones vector takes the form

$$\begin{bmatrix} a_x \\ a_y \end{bmatrix}. \tag{D.6}$$

The normalized form of the Jones vector is obtained by multiplying the full Jones vector by a scalar that reduces the intensity to unity. For light linearly polarized at $45°$, the normalized Jones vector would be

$$2^{-1/2}\begin{bmatrix} 1 \\ 1 \end{bmatrix}. \tag{D.7}$$

Optical elements which modify the state of polarization of the beam are described by matrices containing four elements, known as Jones matrices. For example, the Jones matrix for a half-wave plate, with its axis vertical or horizontal, would be

$$\begin{bmatrix} 1 & 0 \\ 0 & -1 \end{bmatrix}. \tag{D.8}$$

The Jones vector of the incident beam is multiplied by the Jones matrix of the optical element to obtain the Jones vector of the emerging beam. For example, if a beam linearly polarized at 45° is incident on this half-wave plate, the Jones vector for the emerging beam would be

$$\begin{bmatrix} 1 & 0 \\ 0 & -1 \end{bmatrix} 2^{-1/2} \begin{bmatrix} 1 \\ 1 \end{bmatrix} = 2^{-1/2} \begin{bmatrix} 1 \\ -1 \end{bmatrix}, \tag{D.9}$$

which represents a beam linearly polarized at −45°.

In an optical system in which the beam passes through several elements in succession, the overall matrix for the system can be obtained from the Jones matrices of the individual elements by multiplication in the proper sequence.

D.4 THE POINCARÉ SPHERE

The Poincaré sphere is a convenient way of representing the state of polarization of a beam of light; it also makes it very easy to visualize the effects of retarders on the state of polarization of a beam.

As shown in Figure D.2, right- and left-circularly polarized states are represented by the north and south poles of the sphere, while linearly polarized states

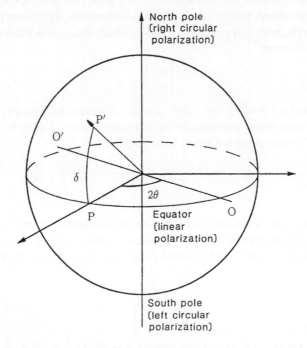

Figure D.2.   The Poincaré sphere; effect of a birefringent plate on the state of polarization of a beam of light.

lie on the equator, with the plane of polarization rotating by 180° for a change in the longitude of 360°. If we consider a point on the equator at longitude 0° as representing a vertical linearly polarized state, any other point on the surface of the sphere with a latitude 2ω and a longitude 2α represents an elliptical vibration with an ellipticity $|\tan\omega|$, whose major axis makes an angle α with the vertical. Two points at the opposite ends of a diameter can be regarded as representing orthogonally polarized states.

A birefringent plate with a retardation δ, whose fast and slow axes are at angles θ and $\theta + 90°$ with the vertical, is represented by the points O and O' at longitudes of 2θ and $2\theta + 180°$. The effect of passage through such a retarder, of a linearly polarized state represented by the point P, is obtained by rotating the sphere about the diameter OO' by an angle δ. The result is that the point P moves to P'.

For more information, see

1. E. Collet, *Polarized Light*, Marcel Dekker, New York (1992).
2. M. Born and E. Wolf, *Principles of Optics*, Cambridge University Press, Cambridge, UK (1999).

E

The Pancharatnam Phase

E.1 THE PANCHARATNAM PHASE

If a beam of light is returned to its original state of polarization *via* two intermediate states of polarization, its phase does not return to its original value but changes by $-\Omega/2$, where Ω is the solid angle (area) spanned on the Poincaré sphere by the geodetic triangle whose vertices represent the three states of polarization. This phase change, known as the Pancharatnam phase, is a manifestation of a very general phenomenon known as the geometric phase.

Unlike the dynamic phase produced by a change in the length of the optical path (for example, by a moving mirror), which is inversely proportional to the wavelength, the geometric phase is a topological phenomenon and, in principle, does not depend on the wavelength. In the case of the Pancharatnam phase, it only depends on the solid angle subtended by the closed path on the Poincaré sphere. As a result, the phase shift produced, even with simple retarders, is very nearly achromatic.

E.2 ACHROMATIC PHASE SHIFTERS

Several systems operating on the Pancharatnam phase can be used as achromatic phase shifters.

With a linearly polarized beam, it is possible to use a combination of a half-wave plate mounted between two quarter-wave plates (a QHQ combination) as an achromatic phase shifter. The two quarter-wave plates have their principal axes fixed at an azimuth of 45°, while the half-wave plate can be rotated.

As shown in Figure E.1, the first quarter-wave plate Q_1 converts this linearly polarized state, represented by the point A_1 on the equator of the Poincaré sphere, to the left-circularly polarized state represented by S, the south pole of the sphere.

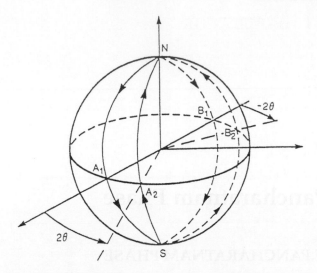

Figure E.1. Poincaré sphere representation of the operation of a QHQ phase shifter operating on the Pancharatnam phase (P. Hariharan and M. Roy, *J. Mod. Opt.* **39**, 1811–1815, 1992).

If, then, the half-wave plate is set with its principal axis at an angle θ to the principal axis of Q_1, it moves this left-circularly polarized state, through an arc that cuts the equator at A_2, to the right-circularly polarized state represented by N, the north pole of the sphere. Finally, the second quarter-wave plate Q_2 brings this right-circularly polarized state back to the original linearly polarized state represented by A_1. Since the input state has been taken around a closed circuit, which subtends a solid angle 4θ at the center of the sphere, this beam acquires a phase shift equal to 2θ.

With two beams polarized in orthogonal planes, the other input state, represented by the point B_1, traverses the circuit $B_1SB_2NB_1$ and acquires a phase shift of -2θ. Accordingly, the phase difference introduced between the two beams is 4θ.

With a Michelson interferometer, it is possible to use a combination of two quarter-wave plates inserted in each beam as an achromatic phase shifter. The quarter-wave plate next to the beam splitter is set with its principal axis at $45°$ to the incident polarization; rotation of the second quarter-wave plate through an angle θ then shifts the phase of the beam by 2θ.

With two orthogonally polarized beams, it is also possible to use a simpler system, consisting of a quarter-wave plate with its principal axis at $45°$ followed by a rotatable linear polarizer, as an achromatic phase shifter. Rotation of the polarizer through an angle θ introduces a phase difference 2θ between the two beams.

E.3 SWITCHABLE ACHROMATIC PHASE SHIFTERS

A QHQ combination can be modified to obtain a switchable, achromatic phase shifter by replacing the half-wave plate with a ferro-electric liquid-crystal (FLC) device.

An FLC device can be regarded as a birefringent plate with a fixed retardation, whose principal axis can take one of two orientations depending on the polarity of an applied voltage. The angle θ through which the principal axis rotates depends on the particular FLC material. Such an FLC device can be used as a binary phase shifter by adjusting its thickness to produce a retardation of half a wave and sandwiching it between two quarter-wave plates. A three-level phase shifter can be constructed using two FLC devices placed between the two quarter-wave plates.

For more information, see

P. Hariharan, *The Geometric Phase*, in Progress in Optics, Vol. XLVIII, Ed. E. Wolf, Elsevier, Amsterdam (2005), pp. 149–201.

F

The Twyman–Green Interferometer: Initial Adjustment

The optical system of a typical Twyman–Green interferometer, with an He–Ne laser source, is shown schematically in Figure F.1. The following step-by-step procedure can be used to set up such an interferometer:

1. Remove the beam-expanding lens (usually, a low-power microscope objective) and the pinhole used to spatially filter the beam, and align the laser

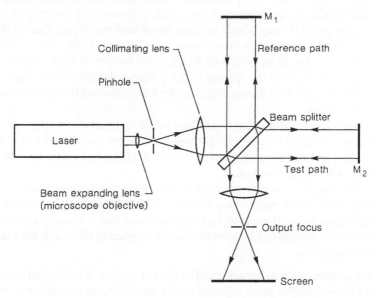

Figure F.1. Optical system of a Twyman–Green interferometer.

187

so that the beam passes through the center of the collimating lens and the beam splitter.

2. Set the end mirrors M_1 and M_2 at approximately the same distance from the beam splitter, and adjust all three so that the two beams are in the same horizontal plane and return along the same paths. This adjustment can be made quite easily if a white card with a 2-mm hole punched at the appropriate height is inserted in each of the beams, first near the end mirrors and then near the beam splitter. When this adjustment is completed, the two beams leaving the interferometer should coincide, and fine interference fringes will be seen if the card is held in the region of overlap.

3. Replace the microscope objective used to expand the laser beam and center it with respect to the laser beam, so that the expanded beam fills the aperture of the collimating lens.

4. Insert the pinhole in its mount and traverse it across the converging beam from the microscope objective. If a white card is held in front of the collimating lens, a bright spot will be seen moving across the card as the pinhole crosses the beam axis. Adjust the pinhole so that this spot is centered on the aperture of the collimating lens.

5. Focus the microscope objective so that the spot expands and fills the aperture of the collimating lens.

6. Focus the collimating lens so that it produces a parallel beam of light. A simple way to perform this adjustment is to view the interference fringes formed on a white card by the light reflected from the two faces of a plane-parallel plate inserted in one beam. With a divergent or convergent beam, straight parallel fringes will be seen, but a uniform, fringe-free field will be obtained when the incident beam is collimated. If the plate is slightly wedged, it should be held with its principal section at right angles to the plane of the beams; the spacing of the fringes will then reach a maximum, and their slope will change sign, as the lens is moved through the correct focus setting.

7. Place a screen with a 1-mm aperture at the focus of the output lens. Two bright, overlapping spots of light will be seen, produced by the two beams in the interferometer, which can be made to coincide by adjusting M_2, the end mirror in the test path.

8. Adjust the position of the aperture to allow these two beams to pass through. Straight, parallel interference fringes will then be seen on a screen placed behind the focus. If the laser beam is attenuated with a neutral filter (density >1.0), the fringes can be viewed directly by placing the eye at the focus of the output lens.

To test an optic, such as a plane-parallel plate or a prism, it is inserted in the test path, and the end mirror M_2 is adjusted so that the test beam is returned along the same path, as shown in Figure 9.4. Two spots of light will then be seen on a card placed in front of the output pinhole and, as before, these spots should be brought into coincidence by adjusting M_2. If necessary, the visibility of the fringes can be optimized by adjusting the length of the reference path.

A problem commonly faced in tests with the Twyman–Green interferometer is deciding whether a curvature of the fringes corresponds to a hill or a valley on the surface of an optic placed in the test path. Where the fringes run across the test optic, a simple solution is to look at the fringes near the edges of the test optic, which, almost always, are slightly rounded off. Where the fringes form closed contours, pressure can be exerted on the base plate of the interferometer, so as to lengthen the test path. If a contour expands, a hill is indicated; conversely, if a contour shrinks, a hollow is indicated.

G

The Mach–Zehnder Interferometer: Initial Adjustment

The mirrors and beam splitters in the Mach–Zehnder interferometer should be provided with tilt adjustments about the horizontal and vertical axes. In addition, the beam splitters should be mounted on micrometer slides, so that they can be moved forward or backward. The following, step-by-step procedure can then be used to adjust the interferometer and obtain white-light interference fringes localized in the test section:

1. Set up a small He–Ne laser, as shown in Figure G.1, so that the beam passes through the center of beam splitter BS_1 and is incident on the center of mirror M_1.

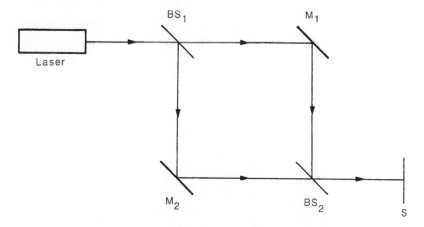

Figure G.1. Adjustment of the Mach–Zehnder interferometer.

2. Adjust beam splitter BS_1, so that the beam reflected by it is incident on the center of mirror M_2.

3. Adjust mirrors M_1 and M_2, so that the beams in opposite arms of the interferometer are parallel. (Check by measurements of their separation at different points along the direction of propagation.)

4. Move beam splitter BS_2, so that the beams from M_1 and M_2 overlap at its center.

5. Adjust beam splitter BS_2, to bring the two spots produced by the output beams on the screen S into coincidence.

6. Repeat step (4), if necessary; then step (5), until the beams are exactly superimposed at BS_2 and S. At this point, interference fringes should be seen in the region of overlap on S. Adjust the beam splitter BS_2 to obtain vertical fringes.

7. Introduce a diffuser, illuminated by a fluorescent lamp, between the laser and the interferometer.

8. Remove the screen S, and view the fringes through a telescope focused on the test section (see Figure 11.1). Adjust the beam splitter BS_2, so that vertical fringes, with a suitable spacing, are obtained.

9. Slowly move beam splitter BS_2 in the direction which reduces the curvature of the fringes, while adjusting beam splitter BS_2 and mirror M_1 to maintain the spacing and improve the visibility of the fringes. The zero-order white fringe (flanked by colored fringes on either side) should appear in the field. If the fringes are lost at any point, go back to step (6).

10. Adjust the beam splitter BS_2 and mirror M_1 to optimize the visibility of the fringes and centralize the zero-order fringe.

Fourier Transforms and Correlation

H.1 FOURIER TRANSFORMS

The Fourier transform of a function $g(x)$ is

$$\mathcal{F}\{g(x)\} = \int_{-\infty}^{\infty} g(x)\exp\left[-i2\pi(\xi x)\right]dx$$
$$= G(\xi),$$ \qquad (H.1)

while the inverse Fourier transform of $G(\xi)$ is

$$\mathcal{F}^{-1}\{G(\xi)\} = \int_{-\infty}^{\infty} G(\xi)\exp\left[-i2\pi(\xi x)\right]d\xi$$
$$= g(x).$$ \qquad (H.2)

These relations can be expressed symbolically in the form

$$g(x) \leftrightarrow G(\xi).$$ \qquad (H.3)

Some useful functions and their Fourier transforms are

Function	Fourier Transform

(i) The delta function

$$\delta(x) = \begin{cases} \infty, & x = 0 \\ 0, & |x| > 0 \end{cases} \qquad 1 \qquad (H.4)$$

Function	**Fourier Transform**

(ii) The rectangle function

$$\text{rect}(x) = \begin{cases} 1, & |x| \leqslant 1/2 \\ 0, & |x| > 1/2 \end{cases} \qquad \text{sinc}(\xi) = (\sin \pi \xi)/\pi \xi \qquad \text{(H.5)}$$

(iii) The circle function

$$\text{circ}(r) = \begin{cases} 1, & r \leqslant 1 \\ 0, & r > 1 \end{cases} \qquad J_1(2\pi\rho)/\rho \qquad \text{(H.6)}$$

H.2 CORRELATION

The cross-correlation of two stationary, random functions $g(t)$ and $h(t)$ is

$$R_{gh}(\tau) = T \overset{\lim}{\to} \infty \frac{1}{2T} \int_{-T}^{T} g(t)h(t+\tau)\, dt, \qquad \text{(H.7)}$$

which can be written symbolically as

$$R_{gh}(\tau) = \langle g(t)h(t+\tau)\rangle. \qquad \text{(H.8)}$$

The autocorrelation of $g(t)$ is, therefore,

$$R_{gg}(\tau) = \langle g(t)g(t+\tau)\rangle. \qquad \text{(H.9)}$$

The power spectrum $S(\omega)$ of $g(t)$ is the Fourier transform of its autocorrelation function

$$S(\omega) \leftrightarrow R_{gg}(\tau). \qquad \text{(H.10)}$$

For more information, see

1. R. M. Bracewell, *The Fourier Transform and Its Applications*, McGraw-Hill, Kogakusha, Tokyo (1978).
2. J. W. Goodman, *Introduction to Fourier Optics*, McGraw-Hill, New York (1996).

I

***

Coherence

Coherence theory is a statistical description of the radiation field due to a light source, in terms of the correlation between the vibrations at different points in the field.

I.1 QUASI-MONOCHROMATIC LIGHT

If we consider a source emitting light with a narrow range of frequencies (a quasi-monochromatic source), the electric field at any point is obtained by integrating Eq. A.1 over all frequencies and can be written as

$$V^{(r)}(t) = \int_0^\infty a(v) \cos\left[2\pi v t - \phi(v)\right] dv. \tag{I.1}$$

For convenience, we define (see Appendix A) a complex function

$$V(t) = \int_0^\infty a(v) \exp\left\{i\left[2\pi v t - \phi(v)\right]\right\} dv, \tag{I.2}$$

which is known as the analytic signal.

The optical intensity due to this quasi-monochromatic source is, then,

$$I = T \overset{\lim}{\to} \infty \frac{1}{2T} \int_{-T}^{T} V(t) V^*(t)\, dt$$

$$= \langle V(t) V^* t \rangle, \tag{I.3}$$

where the pointed brackets denote a time average.

I.2 THE MUTUAL COHERENCE FUNCTION

We can evaluate the degree of correlation between the wave fields at any two points illuminated by an extended quasi-monochromatic point source by the following thought experiment.

As shown in Figure I.1, a quasi-monochromatic source S illuminates a screen containing two pinholes A_1 and A_2, and the light leaving these pinholes produces an interference pattern in the plane of observation.

The wave fields produced by the source S at A_1 and A_2 are represented by the analytic signals $V_1(t)$ and $V_2(t)$, respectively. A_1 and A_2 then act as two secondary sources, so that the wave field at a point P in the interference pattern produced by them can be written as

$$V_P(t) = K_1 V_1(t - t_1) + K_2 V_2(t - t_2), \tag{I.4}$$

where $t_1 = r_1/c$ and $t_2 = r_2/c$ are the times needed for the waves from A_1 and A_2 to travel to P, and K_1 and K_2 are constants determined by the geometry of the system.

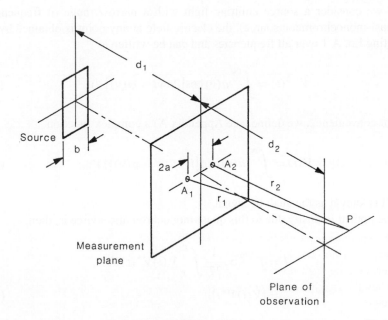

Figure I.1. Measurement of the coherence of the wave field produced by a light source of finite size.

Since the interference field is stationary (independent of the time origin selected), Eq. I.4 can be rewritten as

$$V_P(t) = K_1 V_1(t + \tau) + K_2 V_2(t), \tag{I.5}$$

where $\tau = t_1 - t_2$. The intensity at P is, therefore,

$$
\begin{aligned}
I_P &= \langle V_P(t) V_P^*(t) \rangle \\
&= |K_1|^2 \langle V_1(t + \tau) V_1^*(t + \tau) \rangle + |K_2|^2 \langle V_2(t) V_2^*(t) \rangle \\
&\quad + K_1 K_2^* \langle V_1(t + \tau) V_2^*(t) \rangle + K_1^* K_2 \langle V_1^*(t + \tau) V_2(t) \rangle \\
&= |K_1|^2 I_1 + |K_2|^2 I_2 + 2|K_1 K_2| \mathrm{Re}\{\Gamma_{12}(\tau)\},
\end{aligned}
\tag{I.6}
$$

where I_1 and I_2 are the intensities at A_1 and A_2, respectively, and

$$\Gamma_{12}(\tau) = \langle V_1(t + \tau) V_2^*(t) \rangle \tag{I.7}$$

is known as the mutual coherence function of the wave fields at A_1 and A_2.

I.3 COMPLEX DEGREE OF COHERENCE

Equation I.6 can be rewritten as

$$I_P = I_{P_1} + I_{P_2} + 2(I_{P_1} I_{P_2})^{1/2} \mathrm{Re}\{\gamma_{12}(\tau)\}, \tag{I.8}$$

where $I_{P_1} = |K_1|^2 I_1$ and $I_{P_2} = |K_2|^2 I_2$ are the intensities due to the two pinholes acting separately, and

$$\gamma_{12}(\tau) = \Gamma_{12}(\tau)/(I_1 I_2)^{1/2} \tag{I.9}$$

is called the complex degree of coherence of the wave fields at A_1 and A_2.

I.4 VISIBILITY OF THE INTERFERENCE FRINGES

The spatial variations in intensity observed as P is moved across the plane of observation (the interference fringes) are due to the changes in the value of the last term on the right-hand side of Eq. I.8.

When the two beams have the same intensity, the visibility of the interference fringes is

$$\mathcal{V} = \mathrm{Re}\{\gamma_{12}(\tau)\}. \tag{I.10}$$

The visibility of the interference fringes then gives the degree of coherence of the wave fields at A_1 and A_2.

I.5 SPATIAL COHERENCE

When the difference in the optical paths is small, the visibility of the interference fringes depends only on the spatial coherence of the fields. To evaluate the degree of coherence between the fields at two points P_1 and P_2 illuminated by an extended source S (see Figure I.2), we proceed as follows:

- We first obtain an expression for the mutual coherence of the fields at these two points due to a very small element on the source.
- We then integrate this expression over the whole area of the source.

The resulting expression is similar to the Fresnel–Kirchhoff diffraction integral and leads to the van Cittert–Zernike theorem, which can be stated as follows:

- Imagine that the source is replaced by an aperture with an amplitude transmittance at any point proportional to the intensity at this point in the source.
- Imagine that this aperture is illuminated by a spherical wave converging to a fixed point in the plane of observation (say P_2), and we view the diffraction pattern formed by this wave in the plane of observation.

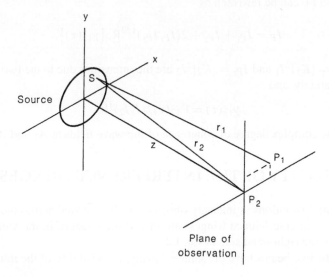

Figure I.2.   Calculation of the coherence of the wave fields at two points illuminated by a light source of finite size.

- The complex degree of coherence between the wave fields at P_2 and some other point P_1 in the plane of observation is then proportional to the complex amplitude at P_1 in the diffraction pattern.

A special case is when the dimensions of the source and the distance of P_1 from P_2 are very small compared to the distances of P_1 and P_2 from the source. The complex degree of coherence of the fields is then given by the normalized two-dimensional Fourier transform (see Appendix H) of the intensity distribution over the source.

I.6 TEMPORAL COHERENCE

For a point source radiating over a range of wavelengths, the complex degree of coherence between the fields at P_1 and P_2 depends only on τ, the difference in the transit times from the source to P_1 and P_2.

The mutual coherence function (see Eq. I.7) then reduces to the autocorrelation function

$$\Gamma_{11}(\tau) = \langle V(t + \tau)V^*(t)\rangle, \tag{I.11}$$

and the degree of temporal coherence of the fields is

$$\gamma_{11}(\tau) = \langle V(t + \tau)V^*(t)\rangle / \langle V(t)V^*(t)\rangle. \tag{I.12}$$

I.7 COHERENCE LENGTH

The frequency spectrum of a source radiating uniformly over a range of frequencies $\Delta \nu$, centered on a mean frequency $\bar{\nu}$, can be written as

$$S(\nu) = \text{rect}\big[(\nu - \bar{\nu})/\Delta \nu\big]. \tag{I.13}$$

Since the mutual coherence function is given by the Fourier transform of the frequency spectrum (see Appendix H), the complex degree of coherence is

$$\gamma_{11}(\tau) = \text{sinc}(\tau \Delta \nu), \tag{I.14}$$

which drops to zero when

$$\tau \Delta \nu = 1. \tag{I.15}$$

The optical path difference at which the interference fringes disappear is

$$\Delta p = c/\Delta \nu \tag{I.16}$$

and is known as the coherence length of the radiation; the narrower the spectral
bandwidth of the radiation, the greater is its coherence length.

For more information, see

M. Born and E. Wolf, *Principles of Optics*, Cambridge University Press, Cambridge, UK (1999).

J

Heterodyne Interferometry

If a frequency difference is introduced between the two beams in an interferometer, the electric fields due to them at any point P can be represented by the relations

$$E_1(t) = a_1 \cos(2\pi v_1 t + \phi_1) \tag{J.1}$$

and

$$E_2(t) = a_2 \cos(2\pi v_2 t + \phi_2), \tag{J.2}$$

where a_1 and a_2 are the amplitudes, v_1 and v_2 the frequencies, and ϕ_1 and ϕ_2 the phases, relative to the origin, of the two waves at the point P. The resultant intensity at P is then

$$\begin{aligned} I(t) &= \left[E_1(t) + E_2(t)\right]^2 \\ &= (a_1^2/2) + (a_2^2/2) \\ &\quad + (1/2)\left[a_1^2 \cos(4\pi v_1 t + \phi_1) + a_2^2 \cos(4\pi v_2 t + \phi_2)\right] \\ &\quad + a_1 a_2 \cos\left[2\pi(v_1 + v_2)t + (\phi_1 + \phi_2)\right] \\ &\quad + a_1 a_2 \cos\left[2\pi(v_1 - v_2)t + (\phi_1 - \phi_2)\right]. \end{aligned} \tag{J.3}$$

The output from a photodetector, which cannot respond to the components at frequencies of $2v_1$, $2v_2$, and $(v_1 + v_2)$ is, therefore,

$$I(t) = I_1 + I_2 + 2(I_1 I_2)^{1/2} \cos\left[2\pi(v_1 - v_2)t + (\phi_1 - \phi_2)\right], \tag{J.4}$$

where $I_1 = (a_1^2/2)$ and $I_2 = (a_2^2/2)$. The phase of the oscillatory component, at the difference frequency $(\nu_1 - \nu_2)$, gives the phase difference between the interfering waves at P, directly.

K

Laser Frequency Shifting

Heterodyne interferometry involves the use of two beams derived from the same laser, one of which has its frequency shifted by a specified amount. A convenient way of introducing such a frequency shift is by means of an acousto-optic modulator.

A typical acousto-optic modulator consists, as shown in Figure K.1, of a glass block with a piezoelectric transducer bonded to it. When the transducer is excited at a frequency ν_m, it sets up acoustic pressure waves which propagate through the glass block, causing a periodic variation (wavelength Λ) in its refractive index. A laser beam (wavelength λ), incident at the Bragg angle θ_B (where $\Lambda \sin\theta_B = \lambda$) on this moving phase grating, is diffracted with a frequency shift ν_m.

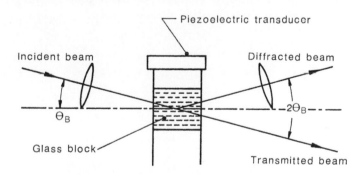

Figure K.1. Frequency shifting of a laser beam by an acousto-optic modulator.

L

Evaluation of Shearing Interferograms

L.1 LATERAL SHEARING INTERFEROMETERS

We consider the case when, as shown in Figure L.1, a shear s along the x-axis is introduced between the two images of the pupil (the test surface) in a lateral shearing interferometer. For convenience, we take the pupil to be a circle of unit diameter. If the wavefront aberration at a point in the pupil with coordinates (x, y) is $W(x, y)$, the optical path difference between the two wavefronts, at the corresponding point in the interferogram, is

$$W'(x, y) = W(x + s/2, y) - W(x - s/2, y). \qquad (L.1)$$

When the shear s is very small, Eq. L.1 reduces to

$$W'(x, y) \approx s[\partial W(x, y)/\partial x]. \qquad (L.2)$$

Since the optical path difference in the interferogram is proportional to the derivative, along the direction of shear, of the wavefront aberration, the errors of the wavefront can be obtained by integrating the values of the optical path difference obtained from two interferograms with mutually perpendicular directions of shear. More accurate measurements can be made by fitting two-dimensional polynomials to the two interferograms. The values of the coefficients of these polynomials are then used to generate a polynomial representing the wavefront aberrations.

L.2 RADIAL SHEARING INTERFEROMETERS

With a radial shearing interferometer, it is convenient to express the aberrations of the test wavefront (see Section 9.3) as a linear combination of circular

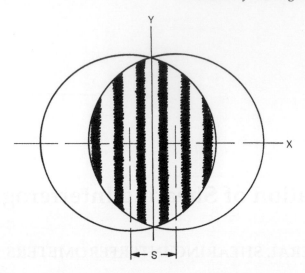

Figure L.1.   Images of the test wavefront in a lateral shearing interferometer.

polynomials in the form

$$W(\rho,\theta) = \sum_{k=0}^{n}\sum_{l=0}^{k} \rho^k (A_{kl} \cos l\theta + B_{kl} \sin l\theta), \qquad (L.3)$$

where ρ and θ are polar coordinates over the pupil (see Figure L.2) and $(k-l)$ is an even number.

If the ratio of the diameters of the two images of the pupil (the shear ratio) is μ, the optical path difference in the interferogram is given by the relation

$$W'(\rho,\theta) = \sum_{k=0}^{n}\sum_{l=0}^{k} \rho^k (A'_{kl} \cos l\theta + B'_{kl} \sin l\theta), \qquad (L.4)$$

where

$$A'_{kl} = A_{kl}\left(1 - \mu^k\right),$$

$$B'_{kl} = B_{kl}\left(1 - \mu^k\right). \qquad (L.5)$$

If the shear ratio is small ($\mu < 0.3$), the interferogram is very similar to that obtained in a Twyman–Green interferometer. For accurate measurements, the wavefront aberrations are evaluated by fitting a polynomial to the interferogram and finding the values of the coefficients A'_{kl} and B'_{kl} in Eq. L.4. Equation L.5 is then

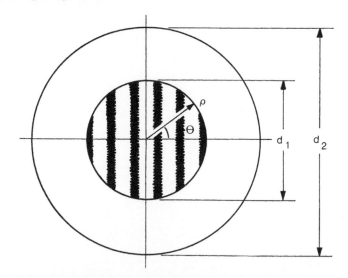

Figure L.2.   Images of the test wavefront in a radial shearing interferometer.

used to calculate the values of the coefficients A_{kl} and B_{kl} in Eq. L.3, for the test wavefront.

For more information, see

1. M. V. Mantravadi, *Lateral Shearing Interferometers*, in Optical Shop Testing, Ed. D. Malacara, John Wiley, New York (1992), pp. 123–172.
2. D. Malacara, *Radial, Rotational and Reversal Shear Interferometers*, in Optical Shop Testing, Ed. D. Malacara, John Wiley, New York (1992), pp. 173–206.

M

Phase-Shifting Interferometry

We assume that the optical path difference between the two beams in an interferometer is changed in steps of a quarter wavelength (equivalent to introducing additional phase differences of 0°, 90°, 180°, and 270°), and the corresponding values of the intensity at each data point in the interference pattern are recorded.

If, then, at any point in the interference pattern, the complex amplitude of the test wave is

$$A = a\exp(-i\phi), \tag{M.1}$$

and that of the reference wave is

$$B = b\exp(-i\phi_R), \tag{M.2}$$

the four values of intensity obtained at this point are

$$I(0) = a^2 + b^2 + 2ab\cos(\phi - \phi_R),$$
$$I(90) = a^2 + b^2 + 2ab\sin(\phi - \phi_R),$$
$$I(180) = a^2 + b^2 - 2ab\cos(\phi - \phi_R),$$
$$I(270) = a^2 + b^2 - 2ab\sin(\phi - \phi_R). \tag{M.3}$$

The phase difference between the test and reference waves, at this point, is then given by the relation

$$\tan(\phi - \phi_R) = \frac{I(90) - I(270)}{I(0) - I(180)}. \tag{M.4}$$

M.1 ERROR-CORRECTING ALGORITHMS

Systematic errors can arise in the values of the phase difference obtained with Eq. M.4 from several causes. The most important of these are: (1) miscalibration of the phase steps, (2) nonlinearity of the photodetector, and (3) deviations of the intensity distribution in the interference fringes from a sinusoid, due to multiply reflected beams. These errors can be minimized by using an algorithm with a larger number of phase steps; the simplest is one using five frames of intensity data recorded with phase steps of 90°. In this case, we have

$$\tan(\phi - \phi_R) = \frac{2[I(90) - I(270)]}{2I(180) - I(360) - I(0)}. \tag{M.5}$$

For more information, see

J. E. Greivenkamp and J. H. Bruning, *Phase Shifting Interferometry*, in Optical Shop Testing, Ed. D. Malacara, John Wiley, New York (1992), pp. 501–598.

N

Holographic Imaging

N.1 HOLOGRAM RECORDING

To record a hologram, a laser is used to illuminate the object and, as shown in Figure N.1, the light scattered by it is allowed to fall directly on a high resolution photographic film. A reference beam derived from the same laser is also incident on the film.

If the reference beam is a collimated beam incident at an angle θ on the photographic film, the complex amplitude due to it, at any point (x, y), is

$$r(x, y) = r \exp(i2\pi\xi x), \tag{N.1}$$

where $\xi = (\sin\theta)/\lambda$, while that due to the object beam is

$$o(x, y) = |o(x, y)| \exp[-i\phi(x, y)]. \tag{N.2}$$

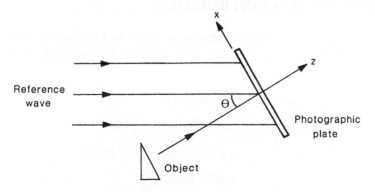

Figure N.1. Hologram recording.

211

The resultant intensity in the interference pattern is, therefore,

$$
\begin{aligned}
I(x, y) &= \left| r(x, y) + o(x, y) \right|^2 \\
&= r^2 + \left| o(x, y) \right|^2 \\
&\quad + r \left| o(x, y) \right| \exp\{-i[2\pi\xi x + \phi(x, y)]\} \\
&\quad + r \left| o(x, y) \right| \exp\{i[2\pi\xi x + \phi(x, y)]\} \\
&= r^2 + \left| o(x, y) \right|^2 + 2r \left| o(x, y) \right| \cos[2\pi\xi x + \phi(x, y)].
\end{aligned} \tag{N.3}
$$

The interference pattern consists of a set of fine fringes with an average spacing $1/\xi$, whose visibility and local spacing are modulated by the amplitude and phase of the object wave.

We will assume that the amplitude transmittance of the film on which the hologram is recorded varies linearly with the exposure and is given by the relation

$$
t = t_0 + \beta T I, \tag{N.4}
$$

where t_0 is the transmittance of the unexposed film, T is the exposure time, and β (negative) is a constant of proportionality. The amplitude transmittance of the hologram is, therefore,

$$
\begin{aligned}
t(x, y) &= t_0 + \beta T \big[r^2 + \left| o(x, y) \right|^2 \\
&\quad + r \left| o(x, y) \right| \exp\{-i[2\pi\xi x + \phi(x, y)]\} \\
&\quad + r \left| o(x, y) \right| \exp\{i[2\pi\xi x + \phi(x, y)]\} \big].
\end{aligned} \tag{N.5}
$$

N.2 IMAGE RECONSTRUCTION

To reconstruct the image, the hologram is illuminated once again, as shown in Figure N.2, with the same reference beam. The complex amplitude of the transmitted wave is then

$$
\begin{aligned}
u(x, y) &= r(x, y) t(x, y) \\
&= \left(t_0 + \beta T r^2 \right) r \exp(i2\pi\xi x) \\
&\quad + \beta T r \left| o(x, y) \right|^2 \exp(i2\pi\xi x) \\
&\quad + \beta T r^2 o(x, y) \\
&\quad + \beta T r^2 o^*(x, y) \exp(i4\pi\xi x).
\end{aligned} \tag{N.6}
$$

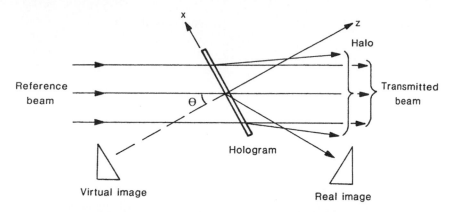

Figure N.2. Image reconstruction by a hologram.

The first term on the right-hand side corresponds to the directly transmitted beam, while the second term gives rise to a halo around it. The third term is similar to the original object wave and produces a virtual image of the object in its original position. The fourth term is the conjugate of the object wave and, as shown in Figure N.2, produces a real image in front of the hologram.

For more information, see

P. Hariharan, *Optical Holography*, Cambridge University Press, Cambridge, UK (1996).

O

Laser Speckle

If an object is illuminated with light from a laser, its image, as shown in Figure O.1, appears covered with a random interference pattern known as a laser speckle pattern. Due to diffraction at the aperture of the imaging lens, the field at any point in the image is the sum of contributions from a number of adjacent points on the object. Since almost any surface is extremely rough on a scale of light wavelengths, the relative phases of these point sources are randomly distributed, though fixed in time. Accordingly, the resultant amplitude varies over a wide range, from point to point in the image, giving rise to a highly irregular interference pattern covering the image.

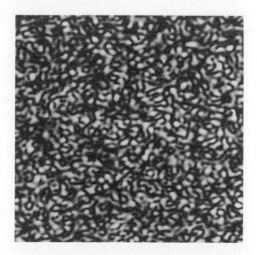

Figure O.1. Speckled image of a rough surface illuminated by a laser.

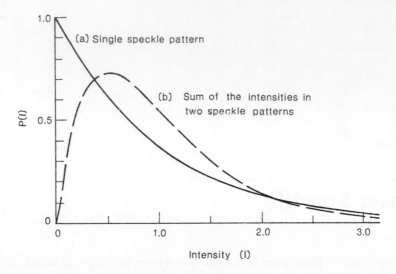

Intensity (I)

Figure O.2. Probability distribution $p(I)$ of (a) the intensity in a single speckle pattern, and (b) the sum of the intensities in two independent speckle patterns.

The average dimensions of the speckles are determined by the aperture of the lens and the wavelength of the light and are

$$\Delta x = \Delta y = 1.22\lambda f/D, \tag{O.1}$$

where f is the focal length of the lens, and D is its diameter, while the probability density of the intensity in the speckle pattern is

$$p(I) = \left(1/\langle I\rangle\right)\exp\left(-I/\langle I\rangle\right), \tag{O.2}$$

where $\langle I\rangle$ is the average intensity. As shown by the solid line in Figure O.2, the most probable intensity is zero, so that there are a large number of completely dark regions, and the contrast of the speckle pattern is very high.

If, however, the images of two independent speckle patterns with equal average intensities are superimposed, the probability density function of the intensity in the resulting speckle pattern is

$$p(I) = \left(4I/\langle I\rangle^2\right)\exp\left(-2I/\langle I\rangle\right), \tag{O.3}$$

which is represented by the broken line in Figure O.2. Since most of the dark regions are eliminated, the contrast of the resulting speckle pattern is much lower.

For more information, see

J. W. Goodman, *Statistical Properties of Laser Speckle Patterns*, in Laser Speckle and Related Phenomena, Topics in Applied Physics, Vol. 9, Ed. J. C. Dainty, Springer-Verlag, Berlin (1975), pp. 9–75.

P

Laser Frequency Modulation

We consider a beam of light (frequency ν, wavelength λ) incident normally on a mirror vibrating with an amplitude a at a frequency f_s. The wave reflected from the mirror then exhibits a time-varying phase modulation

$$\Delta\phi(t) = (4\pi a/\lambda) \sin 2\pi f_s t, \tag{P.1}$$

and the electric field due to the reflected beam can be written as

$$E(t) = E \sin\left[2\pi \nu t + (4\pi a/\lambda) \sin 2\pi f_s t\right]. \tag{P.2}$$

If the vibration amplitude is small, so that $(2\pi a/\lambda) \ll 1$, Eq. P.2 can be written as

$$E(t) \approx E\left[\sin 2\pi \nu t + (2\pi a/\lambda) \sin 2\pi(\nu + f_s)t\right.$$
$$\left. - (2\pi a/\lambda) \sin 2\pi(\nu - f_s)t\right]. \tag{P.3}$$

Reflection at the vibrating mirror generates sidebands at frequencies of $(\nu + f_s)$ and $(\nu - f_s)$. The vibration amplitude can then be determined from a comparison of the components at the original laser frequency and at the sideband frequencies.

This comparison can be made conveniently in the radio-frequency region by interference with a reference beam with a frequency offset. For convenience, we will assume that frequency offsets of $+f_0$ and $-f_0$, respectively, are introduced in the two beams by means of acousto-optic modulators (see Appendix K). The fields due to the two interfering waves can then be written (see Eq. P.2) as

$$E_1(t) = E_1 \sin\left[2\pi(\nu - f_0)t + (4\pi a/\lambda) \sin 2\pi f_s t\right] \tag{P.4}$$

and

$$E_2(t) = E_2 \sin\left[2\pi(\nu + f_0)t + \phi\right], \tag{P.5}$$

where ϕ is the average phase difference between the two waves.

For small vibration amplitudes (see Eq. P.3), the time-varying component observed in the output from a photodetector is then (see Appendix J)

$$\begin{aligned} I(t) \approx E_1 E_2 \{ &\cos(4\pi f_0 t + \phi) \\ &+ (2\pi a/\lambda)\cos\left[(4\pi f_0 - 2\pi f_s)t + \phi\right] \\ &- (2\pi a/\lambda)\cos\left[(4\pi f_0 + 2\pi f_s)t + \phi\right] \}. \end{aligned} \tag{P.6}$$

The vibration amplitude can be evaluated by comparing the power at the sideband frequencies $(2f_0 \pm f_s)$ with that at the offset frequency $2f_0$.

Index

Printed and bound by CPI Group (UK) Ltd, Croydon, CR0 4YY

03/10/2024

01040410-0002